Control and Freedom

Control and Freedom

Power and Paranoia in the Age of Fiber Optics

Wendy Hui Kyong Chun

The MIT Press   Cambridge, Massachusetts   London, England

MIT Press books may be purchased at special quantity discounts for business or sales promotional use. For information, please e-mail ⟨special_sales@ mitpress.mit.edu⟩ or write to Special Sales Department, The MIT Press, 55 Hayward Street, Cambridge, MA 02142.

This book was set in Janson and Rotis Semi Sans on 3B2 by Asco Typesetters, Hong Kong. Printed and bound in the United States of America.

Library of Congress Cataloging-in-Publication Data
Chun, Wendy Hui Kyong, 1969–
  Control and freedom : power and paranoia in the age of fiber optics / by Wendy Hui Kyong Chun.
     p.   cm.
  Includes bibliographical references and index.
  ISBN 0-262-03332-1 (hc : alk. paper)
  1. Optical communications. 2. Fiber optics. 3. Technology and civilization. I. Title.
TK5103.59.C58   2005
303.48′33—dc22   2004061362

10 9 8 7 6 5 4 3 2 1

Contents

Preface

During the Afghanistan War, the second Gulf War, and the subsequent occupation of Iraq, T-shirts, bumper stickers, and politicians reminded us, "Freedom is not free." This phrase, engraved on the Korean War Memorial in Washington, D.C., would seem simply to say that freedom comes at the cost of soldiers' lives and civilian sacrifices. Freedom is not without cost; someone has to pay a price. This phrase, however, is open to another reading: when freedom is conflated with security, freedom loses its meaning—freedom is no longer free. If freedom is reduced to a gated community writ large or becomes the ideological watchword of a national security state, then it can turn into nothing more than the partner of, or the alibi for, control. The very phrase "freedom is not free" can make freedom unfree when it calls on people to accept unfreedom as the cost of freedom. *Free* can also mean priceless, a gift. In English, the word *free* stems from the Sanskrit word for "dear" or "beloved." The phrase "freedom is not free" should never make sense, for what is free should never be devalued. The value of freedom underlined by its etymology is erased when we shift the emphasis away from the action of giving something freely—not in return for something else—to the economism or opportunism of a recipient, looking for a bargain, who refuses to acknowledge this liberality and thus literally cheapens this act. This cheapening of freedom is crucial to the conflation of control with freedom.

Control and Freedom: Power and Paranoia in the Age of Fiber Optics examines "freedom" through the rubric of the Internet, more specifically, through its emergence as a mass medium. Emphasizing the roles of sexuality and race, this book traces the ways in which a technology, which thrives on control, has been accepted, however briefly, as a mass medium of freedom. Moving from utopian narratives about cyberspace to the underlying hardware the Internet seeks to obscure (and about which we often forget), it traces the structuring paradox of information and communications: without control technologies, no freedom (of choice or movement). But the linkage is not an identity: freedom is not the same thing as control. Their conflation is a response to the failures of both liberty and discipline and marks a significant shift in the apparatuses of power: it is a response to the end of the Cold War and to the successes and failures of containment (in Paul Edwards's words, its "closed world"). This conflation of freedom with control also produces and is produced by paranoia, a paranoia that stems from the attempt to solve political problems technologically. To be paranoid is to think like a machine.

In this book, I do not condemn the Internet—if anything, I hold it dear. Liking it or hating it, as such, is as pointless as being "optimistic" or "pessimistic" about its future. Rather, what we need is a serious engagement with the ways in which the Internet enables communications between humans and machines, enables—and stems from—a freedom that cannot be controlled. Because freedom is a fact we all share, we have decisions to make: freedom is not the result of our decisions, but rather, as Friedrich Schelling and Jean Luc Nancy have argued, what makes our decisions possible. This freedom is not inherently good, but entails a decision for "good"—habitation and limitation—or for "evil"—destruction. The gaps within technological control, the differences between technological control and its rhetorical counterpart, and technology's constant failures mean that our control systems can never entirely make these decisions for us.

Fiber-optic networks, this book argues, enable communications that physically instantiate and thus shatter enlightenment; they also link together disparate locations that only sometimes communicate. We must take seriously the vulnerability that comes with communications—not so that we simply condemn or accept all vulnerability without question but so that we might work together to create vulnerable systems with which we can live.

Acknowledgments

I am very grateful to all those who have read and sponsored various parts of this book. I owe special thanks to Diana Fuss and Thomas Keenan, who oversaw this project in its first incarnations so many years and revisions ago; to Bruce Simon, who read and re-read numerous early drafts; to Richard Dienst, Wolfgang Ernst, N. Katherine Hayles, Peter Krapp, Alan Liu, Geert Lovink, Timothy Murray, Ellen Rooney, and Elizabeth Weed, who generously read and commented on later versions of the manuscript; and to Rey Chow, Laura Chrisman, James Der Derian, Madhu Dubey, Amy Kapcynski, Rachel Lee, Nicholas Mirzoeff, Margaret Morse, Lisa Nakamura, Julie Levin Russo, Marc Steinberg, Jeffrey Tucker, and Sau-ling Wong, who all read and offered critiques of portions of it. Their comments have immeasurably improved this book. I have learned much and received great support from my colleagues in the Department of Modern Culture and Media: Nancy Armstrong, Rey Chow, Tony Cokes, Mary Ann Doane, Lynne Joyrich, Roger Mayer, Ellen Rooney, Phil Rosen, Michael Silverman, Len Tennenhouse, and Leslie Thornton; Gayatri Chakravorty Spivak has been an invaluable teacher. Liz Canner, Chris Csikszentmihalyi, Arindam Dutta, Maria Fernandez, and Natalie Jeremijenko are all fellow travelers, to whom I owe inspiration and good cheer. I am also grateful to my wonderful research assistants, Julie Levin Russo, Melanie Kohnen, Yumi Lee, and Michelle Higa, to Joy Simon, who sent me many pertinent articles early on, to Liza Hebert and Susan McNeil (the heart and soul of MCM), as well as the fantastic editorial machine at MIT: Doug Sery, Valerie Geary, and Deborah Cantor-Adams. Martha Fieltsch's and Natalka Migus's unfailing friendship kept me sane during the process of writing. Without the love and support of my partner Paul Moorcroft, this book would not have been possible.

This book is dedicated to my parents Yeong Shik Chun and Soon Jom Chun, in love and respect.

Research for this book was supported by grants, a fellowship and leave from Brown University (in particular a Henry Merritt Wriston Fellowship)—I am grateful to Brown University for its financial and academic support. A fellowship from the Radcliffe Institute for Advanced Study at Harvard University was crucial to completing the manuscript; fellowships from the Social Sciences and Humanities Research Council (Canada) and Princeton University were crucial to its commencement. I am also grateful to Matt Mahurin for freely allowing me to reprint his photographs.

A portion of the introduction has appeared in "Human-Mediated-Communications," *Reality/Simulacra/Artificial: Ontologies of Postmodernity*, in English and Portuguese, ed. Enrique Larreta (Rio de Janeiro: Universidade Candido Mendes, 2003); and a portion of chapter 1 as "Othering Space" *Visual Culture Reader 2.0*, ed. Nick Mirzoeff (New York: Routledge 2003)—both in altered form. An early draft of chapter 3 was published as "Scenes of Empowerment: Virtual Racial Diversity and Digital Divides" *New Formations* 45 (winter 2001); and an early, condensed version of chapter 4 in "Orienting Orientalism, or How to Map Cyberspace," *Asian American.net*, eds. Rachel Lee and Sau-ling Wong (New York: Routledge 2003).

Control and Freedom

INTRODUCTION

We have lived in, and still live in, exciting times, from the fall of the Berlin wall to the heady days of the dot-com era, from the events of September 11, 2001, to the ongoing turmoil in geopolitical relations. All these events have been linked to freedom: the triumph of the Free World, the free market, and the free circulation of information; threats to freedom from abroad, and the U.S. mission to spread democracy and freedom. All these events have also been linked to technology and networks: Eastern Europe's collapse has been attributed to computer technology and broadcast/satellite television; terrorist networks turn everyday technologies like airplanes and cell phones into weapons; the U.S. military's and intelligence agencies' control and communications networks are without rival, if not without fault. But what does it mean to attribute such causality to technology and link freedom to what are essentially control technologies?

Control and Freedom: Power and Paranoia in the Age of Fiber Optics responds to this question by revealing how power now operates through the coupling of control and freedom. Although ideologies and practices of freedom and control are not new, the coupling of these terms is uniquely tied to information technology and our current political situation. Control-freedom, which is intimately experienced as changes in sexuality and race, is a reaction to the increasing privatization of networks, public services and space, and to the corresponding encroachment of publicity and paranoia into everyday life. The end of the Cold War has not dispelled paranoia but rather spread it everywhere: invisibility and uncertainty—of the enemy, of technology—has invalidated deterrence and moved paranoia from the pathological to the logical. This twinning of control and freedom subverts the promise of freedom, turning it from a force that simultaneously breaks bonds and makes relation possible to the dream of a gated community writ large. This subversion of freedom, however, does not forever render freedom innocuous, for if anything cannot be controlled it is freedom. The emergence of the Internet as a mass medium, this book argues, epitomizes this new structure of power and the possibilities for a freedom beyond control.

The Internet as ~~Mass~~ Medium

The Internet, conflated with cyberspace, was sold as a tool of freedom, as a freedom frontier that by its nature could not be tamed: the Internet supposedly interpreted censorship as damage and routed around it.[1] Further, by enabling anonymous communications, it allegedly freed users from the limitations of their bodies, particularly the limitations *stemming* from their race, class, and sex, and more ominously, from social responsibilities and conventions. The Internet also broke media monopolies by enabling the free flow of information, reinvigorating free speech and democracy. It supposedly proved that free markets—in a "friction-free" virtual environment—could solve social and political problems. Although some condemned the Internet for its excessive freedoms, for the ways in which it encouraged so-called deviant behavior that put our future at risk, the majority (of the Supreme Court at least) viewed the Internet as empowering, as creating users rather than couch potatoes, as inspiring Martin Luthers rather than channel surfers.

This rhetoric of the Internet as freedom, excessive or not, was also accompanied by Internet rumors of the Internet as a dark machine of control. For many, Echelon—a shadowy intelligence network operated by the United States, the United Kingdom, Canada, Australia, and New Zealand, and stemming from the 1947 UKUSA agreement in which the Anglo allies turned their antennas from Berlin to Moscow—epitomized the dangers of high-speed telecommunications networks, even though its exact capabilities (especially its ability to penetrate fiber-optic networks) and goals both remain unclear.[2] For others, mysterious corporate "cookies,"

---

1. This phrase is usually attributed to John Gilmore.

2. See Friedrich Kittler's "Cold War Networks or Kaiserstr. 2, Neubabelsberg," in *New Media, Old Media: A History and Theory Reader*, eds. Wendy Hui Kyong Chun and Thomas Keenan (New York: Routledge, 2005), 181–186.

allegedly capable of following our every move, or voracious "packet sniffers" epitomized the risk of going online. The Internet, rather than enabling freedom, enabled total control.

So, was or is the Internet a tool of freedom or control? Does it enable greater self-control or surveillance? *Control and Freedom: Power and Paranoia in the Age of Fiber Optics* argues that these questions and their assumptions are not only misguided but also symptomatic of the increasingly normal paranoid response to and of power. This paranoia stems from the reduction of political problems into technological ones—a reduction that blinds us to the ways in which those very technologies operate and fail to operate. The forms of control the Internet enables are not complete, and the freedom we experience stems from these controls; the forms of freedom the Internet enables stem from our vulnerabilities, from the fact that we do not entirely control our own actions.

Consider, for instance, what happens when you browse a Web page. Your computer sends information, such as your Internet Protocol (IP) address, browser type, language preference, and userdomain (your userdomain often contains information such as your physical location or username).[3] More important, the moment you "jack in" (for networked Macs and Windows machines, the moment you turn on your computer), your Ethernet card participates in an incessant "dialogue" with other networked machines. You can track this exchange using a packet sniffer, a software program that analyzes—that is, stores and represents—traffic traveling through a local area network (see figure 1).[4] Your screen, with its windows and background, suggests that your computer only sends and receives data at your request. It suggests that you are that all-powerful user Microsoft invoked to sell its Internet Explorer by asking, "Where do you want to go today?" Using a packet sniffer, however, you can see that your computer constantly wanders without you. Even when you are not

---

3.   As discussed in more detail in chapter 2, Hypertext Transfer Protocol (HTTP) headers include "from" (your e-mail address), and "Client-IP" (your IP address), and "Referer" (Universal Resource Locator of the document that contains the request Universal Resource Identifier), among many others.

4.   For more on packet sniffers, see the Sniffer FAQ. ⟨http://www .robertgralpubs/sniffing-tag.html⟩ (accessed September 1, 2003).

"using," your computer sends and receives, stores and discards—that is, reads—packets, which mostly ask and respond to the question "Can you read me?" These packets are anything but transparent to you, the user: not only must you install a sniffer to see them; you must also translate them from hexadecimal—that is, if your operating system (OS) allows you to install a sniffer, which classic Macs do not.

Screening this traffic and making analogous browsing the Web and reading a "page" focuses attention on the text and the images pulsing from the screen, rather than on the ways in which you too are coded and circulated numerically, invisibly, nonvolitionally. Rather than simply allowing people to exercise what Walter Benjamin once called their "legitimate claim to be reproduced," the Internet circulates their "reproductions" without their consent and knowledge.[5] Also, rather than simply shattering tradition and bursting open "our prison world," computation's rampant reproductions—its reading as writing elsewhere—literalize control (that is, if it did not make the literal metaphorical). According to the *Oxford English Dictionary*, the English term *control* is based on the French *contreroule*—a copy of a roll of an account and so on, of the same quality and content as the original. This control gives users greater access to each other's reproductions.

Putting sniffers into "promiscuous mode," for instance, accesses all the traffic going through a cable. Depending on the network topology (in older networks bus versus star; in newer ones hub versus switch) and the sniffer's location, the sniffer may access a lot of information or very little. Significantly, though, Ethernet cards routinely read in all packets and then discard those not addressed to it; promiscuous mode does not alter an Ethernet card's normal reading habits. The client-server model of the World Wide Web, in which your computer (the client) only receives data from machines designated as servers, is a software and cultural construction. Every computer with an Ethernet card serves information. This active reading reveals that *for now*, data is cheap and reproducible in ways that defy, rather than support, private property, although those lob-

---

5.   Walter Benjamin, "The Work of Art in the Age of Mechanical Reproduction," in *Illuminations: Essays and Reflections*, trans. Harry Zohn (New York: Schocken Books, 1968), 232.

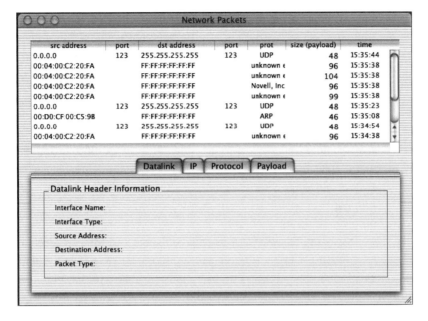

| Figure 1 |
Packet sniffer

bying for stronger copyright laws have also argued that every electronic reading potentially infringes copyright for the same reason. This machine reading makes our digital traces resilient.

Importantly, without this incessant and seemingly disempowering exchange of information, there would be no user interactions, no Internet. The problem is not with the control protocols that drive the Internet—which themselves assume the network's fallibility—but rather with the way these protocols are simultaneously hidden and amplified. This exchange does not inherently enable global surveillance. Fantasies about corporate cookies that malevolently track our every online interaction or unfailing global spy systems also mask the constant, nonvolitional exchange of information that drives the Internet. The Internet as an unfailing surveillance device is thus the obverse, not the opposite, of the Internet as an agency-enhancing marketplace, for it too gives purpose—maps as volitional and permanent—nonvolitional and uncertain software-dependent interactions. This myth also screens the impossibility of storing, accessing, and analyzing everything. Even the U.S. National Security Agency (NSA) admits this impossibility, which is why its intercept equipment automatically stores encrypted packets. The enormous, ever-increasing amount of unanalyzed data belies the computer's analytic promise and demarcates the constitutive boundaries of an "information society." Furthermore, this myth contradicts people's everyday experiences with computers by concealing the ephemerality of information (computer memory is an oxymoron), and the importance of software and local conditions. Computers crash on a regular basis, portable storage devices become unreadable, and e-mail messages disappear into the netherworld of the global network, and yet many people honestly believe in a worldwide surveillance network in which no piece of data is ever lost.

These paranoid narratives of total surveillance and total freedom are the poles of control-freedom, and are symptomatic of a larger shift in power relations from the rubric of discipline and liberty to that of control and freedom.

Control and Freedom

Gilles Deleuze has most influentially described control societies in his "Postscript on Control Societies," in which he argues that we are moving from disciplinary societies, as outlined by Michel Foucault in *Discipline and Punish*, to control societies. According to Foucault, disciplinary societies emerged in the eighteenth century in response to the rise of capitalism and the attendant need for useful bodies. The disciplines offered a finer resolution than sovereign power at a lower cost: the disciplines made power productive, continuous, and cost-effective by moving the emphasis from the body of the king to those "irregular bodies, with their details, their multiple movements, their heterogeneous forces, their spatial relations."[6] Discplinary power differed from sovereign power absolutely: sovereign power was based on the physical existence of the sovereign, who exercised his power spectacularly, if discontinuously. His was a power to inflict death. Disciplinary power operated through visible yet unverifiable apparatuses of power that sought to fabricate individuals through isolation

6. Michel Foucault, *Discipline and Punish: The Birth of the Prison*, trans. Alan Sharing (New York: Vintage Books, 1978), 208.

and constant examination—it was a power over life. Describing the measures taken in response to the plague, Foucault argues, "the enclosed, segmented space, observed at every point, in which the individuals are inserted in a fixed place, in which the slightest movements are supervised, in which all events are recorded, in which an uninterrupted work of writing links the centre and the periphery ... all this constitutes a compact model of the disciplinary mechanism."[7]

The Panopticon encapsulated the disciplinary mechanism for Foucault. Proposed by Jeremy Bentham as a humane and cost-effective solution to dark, festering prisons, unsanitary hospitals, and inefficient schools and workhouses, the Panopticon comprised a central guard tower and a shorter outer annular structure (with windows on the outer circumference and iron gating on the inner) in which the prisoners/workers/patients were individually housed. In the Panopticon, visibility was a trap—the inhabitants could always be viewed by the central tower, but since the windows of the central tower were to be covered by blinds (except during chapel service), they could never be certain when they were being watched. The major effect of the Panopticon was to "induce in the inmate a state of conscious and permanent visibility that assures the automatic functioning of power."[8] To work, power had to be visible, yet unverifiable. Panoptic discipline worked by causing the inmate/worker/student to recreate his or her world, to internalize the light and become light, within an enclosed space.[9] A bourgeois society formally committed to "liberty,

---

7.   Ibid., 197.

8.   Ibid., 201.

9.   Not accidentally, this process of re-creation parallels the process of paranoid recovery. As, to cite Sigmund Freud, "the paranoiac builds [the world] again, not more splendid, it is true, but at least so that he can once more live in it," the inmate/student/worker is called to rebuild their own interior world. If the paranoiac "builds [their world] up by the work of [their] delusion," the inmate/student/worker rebuilds their world by the work of the delusion of constant surveillance. As with the paranoiac, "*the delusion-formation, which we take to be a pathological product, is in reality an attempt at recovery, a process of reconstruction*" (Sigmund Freud, "Psychoanalytic Notes upon an Autobiographical Account of a Case of Paranoia [Dementia Paranoides])," in *Three Case Histories* ([New York: Collier Books, 1963], 147). Rehabilitation becomes paranoid reconstruction.

equality, fraternity" thus needed the disciplines, for as Foucault asserts, the disciplines serve as a sort of "counter-law," introducing asymmetries and excluding reciprocities in a facially equal system. Creating a "private link" between people, the disciplines bring about the nonreversible subordination of one group of people by another, so that "surplus" power is always fixed on the same side.[10]

Deleuze maintains that the confinement and the mass individuation symptomatic of disciplinary societies is now yielding to flexibility and codes—that is, control. Control society is not necessarily better or worse than disciplinary society; rather, it introduces new liberating and enslaving forces. Whereas disciplinary society relied on independent variables or molds, control society thrives on inseparable variations and modulations: factories have given way to businesses with "souls" focused on metaproduction and on destroying unions through inexorable rivalry; schools have given way to continuing education and constant assessment; new prison techniques simultaneously offer greater freedom of movement and more precise tracking; and the "new medicine 'without doctors and

---

10.   For Foucault, power is not something that one possesses, nor is it a force that simply represses. Rather, as he argues in *The History of Sexuality, Volume I: An Introduction*, trans. Robert Hurley (New York: Vintage Books, 1978):

Power must be understood in the first instance as the multiplicity of force relations immanent in the sphere in which they operate and which constitute their own organization; as the process which, through ceaseless struggles and confrontations, transforms, strengthens, or reverses them; as the support which these force relations find in one another, thus forming a chain or a system, or on the contrary, the disjunctions and contradictions which isolate them from one another; and lastly, as the strategies in which they take effect, whose general design or institutional crystallization is embodied in the state apparatus, in the formulation of the law, in various social hegemonies ... it is the moving substrate of force relations which by virtue of their inequality, constantly engender states of power, but the latter are always local and unstable. (92–93).

Power is not something that exists abstractly, but only exists in its application; also, where there is power, there is resistance. Importantly, as he argues in "Two Lectures" (in *Power/Knowledge: Selected Interviews and Other Writings, 1972–1977*, ed. Colin Gordon [New York, Pantheon Books, 1980, 78–108]) the fact that power exists in and creates a net-like structure in which everybody acts does not mean "power is the best distributed thing in the world, although in some sense that is so. We are not dealing with a sort of democratic or anarchic distribution of power through bodies" (99).

patients' identifies potential cases and subjects at risk" without attempting treatment. According to Deleuze, these all "form a system of varying geometry whose language is *digital* (though not necessarily binary)."[11] The computer, with its emphasis on information and its reduction of the individual to the password, epitomizes control societies. Digital language makes control systems invisible: we no longer experience the visible yet unverifiable gaze but a network of nonvisualizable digital control.

Deleuze's reading of control societies is persuasive, although arguably paranoid, because it accepts propaganda as technological reality, and conflates possibility with probability. Just as panopticism overestimated the power of publicity, so too does control-freedom overestimate the power of control systems.[12] This is not to say that Deleuze's analysis is not correct but rather that it—like so many other analyses of technology—unintentionally fulfills the aims of control by imaginatively ascribing to control power that it does not yet have and by erasing its failures. Thus, in order to understand control-freedom, we need to insist on the failures and the actual operations of technology. We also need to understand the difference between freedom and liberty since control, though important, is only half of the story.

Although used interchangeably, freedom and liberty have significantly different etymologies and histories. According to the *Oxford English Dictionary*, the Old English *frei* (derived from Sanskrit) meant dear and described all those close or related to the head of the family (hence friends). Conversely in Latin, *libertas* denoted the legal state of being free versus enslaved and was later extended to children (*liberi*), meaning literally the free members of the household. Those who are one's friends are free; those who are not are slaves. But, like love, freedom exceeds the subject. Liberty is linked to human subjectivity; freedom is not. The Declaration of Independence, for example, describes men as having liberty and

---

11.   Gilles Deleuze, "Postscript on Control Societies," in *CTRL [SPACE]: Rhetorics of Surveillance from Bentham to Big Brother*, eds. Thomas Y. Levin et al. (Cambridge: MIT Press, 2002), 320–321, 318.

12.   For more on Jeremy Bentham's overestimation of publicity, see Foucault's discussion of the importance of media in "The Eye of Power," in *Power/Knowledge: Selected Interviews and Other Writings, 1972–1977*, 146–165.

the nation as being free. Free will—"the quality of being free from the control of fate or necessity"—may first have been attributed to human will, but Newtonian physics attributes freedom—degrees of freedom, free bodies—to objects.

Freedom differs from liberty as control differs from discipline. Liberty, like discipline, is linked to institutions and political parties, whether liberal or libertarian; freedom is not. Although freedom can work for or against institutions, it is not bound to them—it travels through unofficial networks. To have liberty is to be liberated from something; to be free is to be self-determining, autonomous. Freedom can or cannot exist within a state of liberty: one can be liberated yet "unfree," or "free" yet enslaved (Orlando Patterson has argued in *Freedom: Freedom in the Making of Western Culture* that freedom arose from the yearnings of slaves). Freedom implies—or perhaps has become reduced to—freedom of movement: you drive on a freeway, not a libertyway. Free love and free speech move from location to location, person to person. Hackers declare that information, which is technically a measure of the degree of freedom within a system, should be free. Freedom, in its current distinction from liberty, responds to liberty's inadequacies. Freedom, as freedom of movement, cannot easily endorse segregation—there can be no equal but separate. The "freedom rides" of the civil rights movement responded to emancipation's inadequacies. Crucially, this difference between freedom and liberty makes sense mainly in Anglo languages. U.S. politics, from segregation to late-twentieth- and early-twenty-first-century U.S. global power, arguably generates the pronounced distinction between the two.

In an odd extension of commodity fetishism, we now wish to be as free as our commodities: by freeing markets, we free ourselves.[13] And

---

13.   According to Karl Marx, "The mysterious character of the commodity-form consists … in the fact that the commodity reflects the social characteristics of men's own labor as objective characteristics of the products of labor themselves, as the socio-natural properties of these things.... [I]t is nothing but the definite social relation between men which assumes here, for them, the fantastic form of a relation between things" (*Capital*, vol. 1 trans. Ben Fowkes, [New York: Penguin Books with New Left Review, 1976], 164–166). The commodity now seems to be endowed with *freedom*, operating in a free marketplace: now the desire is to emulate such a commodity.

this freedom is supposed to resonate with all the greatness of prior liberations. If once "white man's burden," it is now "enduring freedom"; if once "liberty, equality, and fraternity," now "freedom, democracy, free enterprise." George W. Bush's new tripartite motto hijacks the civil rights movement, erases equality and fraternity, and makes ambiguous the subject of freedom. Bush asserts that "the concept of 'free trade' arose as a moral principle even before it became a pillar of economics. If you can make something that others value, you should be able to sell it to them. If others make something that you value, you should be able to buy it. This is real freedom, the freedom for a person—or a nation—to make a living."[14] His statement unashamedly and uncannily resonates with Karl Marx's condemnation of bourgeois freedom: "In a bourgeois society capital is independent and has individuality, while the living person is dependent and has no individuality.... By freedom is meant, under the present bourgeois conditions of production, free trade, free selling and buying."[15] Freedom as stemming from a commodity's "natural" qualities reflects capitalism's naturalization and the new (rhetoric of) transparency.

Sexuality in the Age of Fiber Optics

As the rest of this book elaborates, the relationship between control and freedom in terms of fiber-optic networks is often experienced as sexuality or is mapped in terms of sexuality-paranoia.

The insight that power can be experienced as sexuality is indebted to the work of Foucault and the psychotic Daniel Paul Schreber (and Eric Santner's interpretation of his memoirs). Foucault, in the first volume of his uncompleted *History of Sexuality*, contends that sexuality is "the secret" instrumental to power/knowledge. Since modernity, we have constantly confessed the truth of sex: from seventeenth-century Catholic confessions that demanded more and more technical details to 1960s' declarations of sexual freedom and revolt; from psychoanalysis to institutional

14.   Office of the White House, *National Security Strategy of the United States of America*, 〈http://www.whitehouse.gov/nsc/nss.html〉 (accessed October 1, 2003).

15.   Karl Marx and Frederick Engels, *Communist Manifesto* (Peking: Foreign Languages Press, 1975), 52.

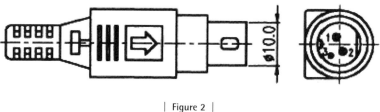

| Figure 2 |
Male connector

architecture. Sexuality is key to determining the subject—its causality, its unconscious, the truth it holds unbeknownst to itself. Sexuality is the meeting point between the two objects of biopower (the power over life): the individual and the species. As such, sexuality is intimately linked to twentieth-century racism (state-sponsored programs to further the survival of the species). Sexuality, for Foucault, is a dense transfer point for relations of power "between men and women, young people and old people, parents and offspring, teachers and students, priests and laity." It "require[s] the social body as a whole, and virtually all of its individuals, to place themselves under surveillance."[16]

Given Foucault's thesis perhaps it is not surprising that sex and sexuality dominate descriptions and negotiations of the thrills and the dangers of networked contact. In terms of hardware, male-to-female connectors configure all electronic information exchange as electrifying heterosexual intercourse (see figures 2, 3, and 4). In terms of software, computer viruses spread like sexually transmitted diseases, contaminating and reproducing uncontrollably.[17] In terms of content, pornography is "all over the Internet," saturating the digital landscape and ranking among its more popular recreational uses. In terms of technology development, sex allegedly popularizes new devices: pornography is the "killer application" that

---

16.   Foucault, *The History of Sexuality, Volume I: An Introduction*, 103, 116.

17.   "Clit.exe" is a command line utility that converts an encrypted Lit book to Hypertext Markup Language (HTML), text, or any other format. In terms of operating systems, the UNIX "finger" command retrieves information about someone's online activities, and one "mounts" a disk.

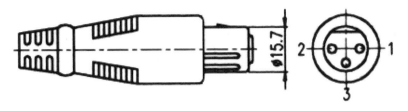

| Figure 3 |
Female connector

convinces consumers to invest in new hardware. New technology is a
"carrier"—a new Trojan horse—for pornography; sex is "a virus that almost always infects new technology first."[18] Sexuality is the linchpin for
strategies as diverse as entrepreneurial capitalism, censorship, and surveillance. Cyberporn fueled the dot-com craze. In terms of censorship and
surveillance, sexuality encapsulated and sequestered, and still encapsulates
and sequesters, the risk of being online; anxiety over or desire for online
contact is expressed as anxiety over or desire for sexual exposure. Before
September 11, 2001, those seeking to censor the Internet, through public
or private means, claimed without fail to be protecting children from the
seamier sides of human sexuality. In the face of catastrophic, unrestrained,
and unrestrainable contact that could compromise our species' fitness, we
were, and are, called to place ourselves under surveillance. Spun more positively, the release of "the seamier sides of human sexuality" encapsulates
the freedom from history or materiality that the Internet promises. This
freedom, however, as Mimi Nguyen has argued, must be read against the
"bodies of Asian and Asian American immigrant women workers (in
sweatshops and factories of varying working conditions) [that] provide
the labor for the production of … circuit boards, those instruments
of identity play, mobility, and freedom."[19] The current explosion in

---

18.   Gerard Van Der Leun, quoted in Mark Dery, *Escape Velocity* (New York:
Grooe Press, 1996), 218.

19.   Mimi Nguyen, "Queer Cyborgs and New Mutants," in *Asian America.Net*,
eds. Rachel Lee and Sau-ling Wong (New York: Routledge, 2003), 300.

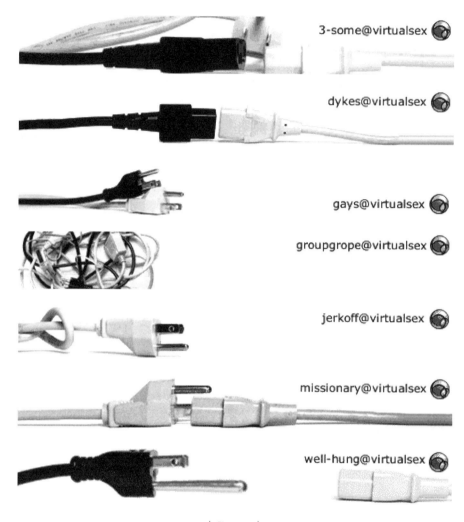

| Figure 4 |
DM9 DDB Publicidade banner ad campaign for Brazilian Internet service provider, UOL-Universo Online,
⟨http://duplo.org/wille/⟩

discourses about sex and sexuality, this book argues, is symptomatic of larger changes in biopower, and is intimately linked to changes in our understanding of race and changes to racism. The relationship between the individual and the species is changing, and the grid of liberties and discipline, which Foucault saw as key to modern power, is malfunctioning, for fiber-optic networks threaten a freedom and a democratization that threaten to verge out of control as well as calls for security bent on destroying them.

The current configuration of fiber-optic networks challenges disciplinary and regulatory power. Telecommunications monopolies, rules, and regulations have been and continue to be revised, many regulatory techniques have been rendered ineffectual, and many new, more invasive techniques are being introduced. The sheer number of Web sites, the multiple paths, and the rapidity with which sites are altered, built, destroyed, and mirrored makes regulation of this new ~~mass~~ medium far more difficult than any other (its closest predecessor is the telephone, which does not broadcast). However, unlike the telephone, it does make prosecution easier: if log files have been cached, one can track visits to a certain Web site or the sending location of e-mails (and one does not need a warrant in the United States or the United Kingdom to access these locations). Prosecution is also easier postevent because by then the search terms are obvious. In addition, the illusion of privacy—the illusion that what one does in front of one's computer in the privacy of one's own home is private—troubles the effectiveness of public standards.

Fiber-optic networks open the home. As Thomas Keenan has argued, all windows both separate and breach public and private spaces: behind the window, one is a knowing subject; before it, a subject "assumes public rights and responsibilities, appears, acts, intervenes in the sphere it shares with other subjects"; but the glaring light that comes through the window—exposing us to others, even before there is an us—is also something soft that breaks.[20] The computer window seems irreparable and unpluggable. In contrast to its predecessors, the jacked-in computer

---

20. Thomas Keenan, "Windows: Of Vulnerability," in *The Phantom Public Sphere*, ed. Bruce Robbins (Minneapolis: University of Minnesota Press, 1997), 132.

———

window melts the glass and molds it into a nontransparent and tentacling cable. If "the philosophical history of the subject or the human is that of a light and a look, of the privilege of seeing and the light that makes it possible," the light that facilitates the look can no longer be seen; we no longer see through the glass that connects, separates, and breaks.[21] Fiber-optic networks enable uncontrollable circulation. Richard Dienst, adapting Martin Heidegger's "Age of the World Picture" to a "theory after television," claims that "caught in the act of representing themselves to themselves ... modern subjects place themselves in the 'open circle of the representable,'" in a "shared and public representation." A subject is thus "what can or believes it can offer itself representations," "formed by the imperative to be an image, in order to receive images."[22] Fiber optics threaten an infinite open circle of the "representable"—they melt and stretch the glass so that nothing screens the subject from the circulation and proliferation of images. At the same time, they displace representation by code, for if Heidegger emphasized representation as a placing before, no "thing" is placed before oneself.[23] Although medical fiber-optics are still looking glasses, fiber-optic networks use glass to relay light pulses that must be translated into voltages: rather than magnifying images, they relay data in a nonindexical manner.

Beyond, Before, in Front of the Screen

To understand control and freedom in the context of fiber-optic networks, this book examines all four layers of networked media—hardware, software, interface, and extramedial representation (the representation of networked media in other media and/or its functioning in larger economic and political systems)—as well as the disconnect between them, and the possibilities and limitations for actions opened by them. It takes up N.

---

21.   Ibid., 110.

22.   Richard Dienst, *Still Life in Real Time: Theory after Television* (Durham, NC: Duke University Press, 1994), 140.

23.   For more on fiber optics, see Jeff Hecht, *City of Light: The Story of Fiber Optics* (New York: Oxford University Press, 1999), and Joseph C. Palais, *Fiber Optic Communications*, 4th ed. (Upper Saddle River, NJ: Prentice Hall, 1998).

Katherine Hayles's call for medium specific criticism, engaging visual *and* nonvisual aspects of networked machines—human and machine readings—as well as their economic and political impact. These aspects taken together reveal the erasures necessary for the Internet's emergence as a ~~mass~~ medium and the possibilities opened by high-speed communications networks for something like democracy.[24]

By engaging the four layers of networked media, this book seeks to mediate between visual culture studies and media archaeology; to exaggerate slightly, the screen divides new media studies into these two fields. Visual culture studies stem from the Anglo-speaking academy and generally treats the interface, or representations of the interface, as the medium. The second approach, media archaeology, although inspired by Marshall McLuhan and Foucault, is mainly Germanic (most specifically, it emerges from the "Sophienstraße" departments of Humboldt University in Berlin). Taking as its ground zero McLuhan's mantras of "the medium is the message" and "the content of a medium is always another medium," media archaeology concentrates on the machine and often ignores the screen's content. Archaeological studies critique visual culture studies' conflation of interface with medium and representation with actuality; visual culture studies critique the archaeologists' technological determinism and blindness to content and the media industry.[25]

This division between visual culture and media archaeology is not set in stone: many in both fields use the same theoretical sources, such as Foucault and Jacques Lacan. As well, many analyses can work both sides of the screen. For example, Lev Manovich's *The Language of New Media* simultaneously investigates the parallels between cinematic and new media history and argues for the emergence of "software studies." His five principles of new media (numerical representation, modularity, automation, variability, and transcoding) enable a formalist understanding of new media with an important twist. His last principle, transcoding, encapsulates his theoretical intervention succinctly: because new media objects are

---

24.   See N. Katherine Hayles, *Writing Machines* (Cambridge: MIT Press, 2002).

25.   For more on the distinction and relationship between the two fields, see Chun, "Introduction: Did Somebody Say New Media?" in *New Media, Old Media: A History and Theory Reader*, 1–10.

---

designed to both make sense to human users and follow established computer conventions, all new media objects consist of two layers (the cultural and the computer). For Manovich, these two layers are not equal: he asserts that media studies must be transformed into software studies. This privileging of software allows Manovich to translate between the unseen and the seen, which is theoretically if not practically possible (one cannot easily read compiled programs). The problem with "software studies" or transcoding, however, is this privileging of software as readable text; it ignores the significance of hardware and extramedial representation because it only moves between software and interface. Also, this notion of transcoding perpetuates the idea that software merely translates between what you see and what you cannot see, effectively erasing the many ways in which they do not correspond.[26]

This emphasis on software repeats the founding gesture of the Internet: the Internet seeks to make irrelevant hardware differences—its protocol enables networks to communicate regardless of which network (IEEE 802.x) standard is being used. Yet software, at a fundamental level, does not exist. As Friedrich Kittler argues, there is no software:

Not only no program, but no underlying microprocessor system could ever start without the rather incredible autobooting faculty of some elementary functions that, for safety's sake, are burned into silicon and thus form part of the hardware. Any transformation of matter from entropy to information, from a million sleeping transistors into differences between electronic potentials, necessarily presupposes a material event called "reset".

In principle, this kind of descent from software to hardware, from higher to lower levels of observation, could be continued over more and more decades. All code operations, despite their metaphoric faculties such as "call" or "return", come down to absolutely local string manipulations and that is, I am afraid, to signifiers of voltage differences.[27]

---

26.   For more on this, see Wendy Hui Kyong Chun, "On Software, or the Persistence of Visual Knowledge," *grey room* 18 (winter 2005), 26–51.

27.   Friedrich Kittler, "There Is No Software" ⟨http://www.ctheory.net/textfile.asp?pick-74⟩ (accessed August 1, 2004).

User control dwindles as one moves down the software stack; software it-self dwindles since everything reduces to voltage differences as signifiers. Although one codes software and, by using another software program, reads noncompiled code, one cannot see software. Software cannot be physically separated from hardware, only ideologically.[28] The term *digital media* stresses hardware, for switches and vacuum tubes determined the difference between analog and discrete computation. Software has no in-trinsic value, and the concept of software itself has changed over time. As Eben Moglen notes in "Anarchism Triumphant: Free Software and the Death of Copyright," any part of a computer configuration that could be altered was initially called software.

Kittler, finessing his statement slightly, states that there would be no software if computer systems were not surrounded by "an environment of everyday languages, everyday languages of letters and coins, books and bucks."[29] Whereas Kittler's brilliant antihumanist critique focuses on humans as bottlenecks to the machinic symbolic system, this book's cri-tique dwells on the persistence of human reading, on the persistence of software as an ideological phenomenon, or to be more precise, as a phe-nomenon that mimics or simulates ideology.

In a *formal* sense, computers understood as comprising software and hardware are ideology machines. They fulfill almost every formal defini-tion of ideology we have, thus revealing the paucity of our understanding of ideology. Consider, for instance, the commonsense (Marxist) notion of ideology as false consciousness, as some false interpretative apparatus that veils one's vision, but that can be torn asunder. The movie *The Matrix* expresses this view succinctly. In *The Matrix*, humans are literally duped by software; software produces an insidious "residual self-image" (a kind of false consciousness) that prevents humans from seeing the real, which is (á la Jean Baudrillard) a desert. Not coincidently, *The Matrix* is a filmic representation, for only cinema could visualize digital media as false

---

28. Those seeking to archive software programs face this indivisibility all the time. Many old software programs cannot be run on current computers, although custom-built virtual computer simulations can get around this difficulty.

29. Kittler, "No Software."

consciousness so compellingly. Through this representation, cinema displaces its own metaphoric relationship to ideology and Plato's cave.[30]

To accept cinema's imaginings as accurate, however, is to philosophize in the dark. In terms of actual computer interfaces, Louis Althusser's definition of ideology as "a 'representation' of the imaginary relation of individuals to their real conditions of existence" resonates most strongly.[31] Software, or perhaps more precisely OS, offer us an imaginary relationship to our hardware: they do not represent the motherboard or other electronic devices but rather desktops, files, and recycling bins. Without OS, there would be no access to hardware—there would be no actions, no practices, no users. Each OS, in its extramedial advertisements, interpellates a "user": calls it and offers it a name or an image with which to identify. So, Mac users "think different" and identify with Martin Luther King and Albert Einstein; Linux users are open-source power geeks, drawn to the image of a fat, sated penguin; and Windows users are mainstream, functionalist types comforted, as Moglen contends, by their regularly crashing computers.[32] Importantly, the "choices" operating systems offer limit the visible and the invisible, the imaginable and the unimaginable. UNIX allows you to have multiple desktops and to share them—as of 2005, neither MacOS nor Windows does this. The only place Microsoft allows you to move its desktop Internet Explorer icon is the trash. You are not, however, aware of software's constant constriction and interpellation (also known as its user-friendliness) unless you find yourself frustrated with its defaults, which are rather remarkably referred to as *your* preferences, or if you use multiple operating systems or competing software packages. The term *user-friendly*, as Natalie Jeremijenko has argued,

---

30. For more on this, see John-Louis Baudry, "The Apparatus: Metapsychological Approaches to the Impression of Reality in the Cinema," in *Narrative, Apparatus, Ideology: A Film Theory Reader*, ed. Philip Rosen (New York: Columbia University Press, 1986), 299–318.

31. Louis Althusser, "Ideology and Ideological State Apparatuses (Notes towards an Investigation)," in *Lenin and Philosophy and Other Essays*, trans. Ben Brewster (New York: Monthly Review Press, 1971), 162.

32. For more on the significance of fat penguins, see Linus Torvalds, "Why a Penguin?" ⟨http://www.linux.org/info/penguin.html⟩ (accessed January 1, 2004).

implies that human users are inert and interchangeable, and that software is active and animate.[33] Of course, users know very well that their folders and recycling bins are not really folders and recycling bins. This knowledge, however, rather than disqualifying the relationship between software and ideology, buttresses it. As Slavoj Žižek, drawing from Peter Sloterdijk, argues, "ideology's dominant mode of functioning is cynical ... 'they know very well what they are doing, but still, they are doing it.[34]'" It is through this continual doing—this "using," this externalization of our beliefs onto objects that act for us—that ideology operates.

Software produces users, and the term *user*, resonating with "drug user," discloses every programmer's dream: to create an addictive product.[35] Users are produced by benign software interactions, from reassuring sounds that signify that a file has been saved to folder names such as "my documents" that stress personal computer ownership. Computer programs shamelessly use shifters, pronouns like "my" and "you," that address you, and everyone else, as a subject. As Margaret Morse has asserted, these shifters are key to post-televisual interactivity, to the emergence of cyberculture (versus information), and to the delegation of "soft" social control to machines.[36] Software makes you read; it offers you more relationships and ever-more visuals. Software provokes readings that go beyond the reading of letters toward the nonliterary and archaic practices of guessing, interpreting, counting, and repeating. If you believe that your communications are private, it is because software corporations, as they relentlessly code and circulate you, tell you that you are behind, and not

---

33. See Natalie Jeremijenko, "Dialogue with a Monologue: Voice Chips and the Products of Abstract Speech," ⟨http://cat.nyu.edu/natalie/VoiceChips.pdf⟩ (accessed September 13, 2002).

34. Slavoj Žižek, *The Sublime Object of Ideology* (London: Verso, 1989), 29.

35. For more on addiction and technology, see Avital Ronell, *Crack Wars: Literature, Addiction, Mania* (Lincoln: University of Nebraska Press, 1992); and Ann Weinstone, "Welcome to the Pharmacy: Addiction, Transcendence, and Virtual Reality," *diacritics* 27, no. 3 (1997): 77–89.

36. "Chapter One: Virtualities" in Margaret Morse, *Virtualities: Television, Media Art, and Cyberculture* (Bloomington: Indiana University Press, 1998), 3–35.

in front of, the window.[37] Even when "lurking," you constantly send information. It is impossible to resist subjectivity by doing nothing (as Baudrillard once argued and encouraged) if we jack in or are jacked in.

Software and ideology seem to fit each other perfectly because both try to map the material effects of the immaterial and posit the immaterial through visible cues. Software's uncanny paralleling of ideology not only reveals its programmers' dreams but also its struggle to emerge as a commodity, as a value. When software programs first emerged, it was unclear that something so insubstantial should be bought or sold. Software's popularity as a heuristic, coupled with the multibillion dollar industry it supports, testifies to its success.[38] As Moglen notes, "The division between hardware and software ... has become a new way to express the conflict between ideas of determinism and free will, nature and nurture, or genes and culture. Our 'hardware,' genetically wired, is our nature, and determines us. Our nurture is 'software,' establishes our cultural programming, which is our comparative freedom," and thus conversely our exposure to control.[39] Although nature as hardware seems to treat nature as inflexible (genetically wired), and therefore lends hardware and networking protocols an undeserved stability and reality, it also makes nature an object of choice, as easily manipulated and upgraded as hardware. This parallel between software and ideology, however, flattens ideology to its similarities to software, and elides the difference between software as code and software as executed program. More important, it suppresses the question of power and struggle, central to any serious study of ideology. Insisting on software as ideology par excellence excellently drains ideology of meaning and reduces it to acts of programming, which can be reprogrammed by

37.   For more on windows and political theory, see Keenan, "Windows."

38.   Again, Lev Manovich's groundbreaking, insightful, and important work, *The Language of New Media*, in its move toward computer science terminology and its call for a move from media theory to software theory (48), perpetuates the dominance of software.

39.   Eben Moglen, "Anarchism Triumphant: Free Software and the Death of Copyright," *First Monday* 4, no. 8 (August 2, 1999), ⟨http://firstmonday.org/issues/issue4_8/moglen/index.html⟩ (accessed May 1, 2004).

individuals-cum-hackers (this is the libertarian message of *The Matrix*). As well, this analysis reveals the many slippages between software, interface, and extramedial representation that must happen in order for software to gain such power.

Thus, against the recent trends in new media studies to view new media as the coming together of computation and media, and to downplay the significance of utopian and dystopian imaginings of cyberspace, this book insists on the importance of extramedial representation, for the Internet (as cyberspace) existed within the public's imagination before it became a regular public practice.[40] The Internet was sold as "theory come true," as the future in the present arrived as expected. During the 1990s, inflated promises, usually accompanied by knowing disappointment, sold the Internet. Much scholarly work, for instance, claimed that electronic texts literalized a theoretical ideal. Influential early work on hypertext argued that it epitomized Roland Barthes's writerly text; almost every major museum claimed and still claims that its Web site is André Malraux's museum without walls. Rhizome.org is/was one of the most influential net art sites, and Critical Art Ensemble, among many others, maintains that the Internet is rhizomic (the Internet has resuscitated Gilles Deleuze and Felix Guattari within North American theory and art circles: doubtless we are all bodies without organs). Early analyses of MOOs and MUDs argued that cybersex cemented Foucault's claim that sexuality is becoming discursive and every angst-ridden boy passing as the girl of

---

40. The disconnect between literary and filmic representations of high-speed telecommunications networks preceding the mass adoption of the Internet and the Internet as we now know it is stark, and many an analysis has foundered by conflating the two. Still, dismissing these influential representations of the future in order to concentrate on the present, as Manovich does in *The Language of New Media*, can also make us founder. Manovich's insistence on the present perpetuates a new rationalism (intensified and sanctified by the dot-bombs), which dismisses and is embarrassed by utopian rhetoric and early Net criticism. To its credit, *The Language of New Media* has moved theory away from virtual reality and William Gibson's cyberspace to the Internet and current computer art practices. Yet Manovich's critique begs the question, Why did so many theoretical and popular speculations on new media erase the difference between the present and the future?

your dreams proved that gender is performative.[41] Popular analyses portrayed the Internet as finally solving the problems of X by fulfilling X's promise. As I argue in chapter 3, almost every television commercial advertising the Internet in the mid- to late 1990s alleged that it substantiated (finally) a marketplace of ideas by eradicating all physical markers of difference. Accordingly, this erasure eradicated the discrimination that supposedly *stemmed from* these differences. Al Gore declared that the Internet was a revitalized Greek agora. Bill Gates claimed that the Internet was a space for "friction-free capitalism." The Internet was a global village, albeit a happier one than McLuhan's. Even to call the Internet cyberspace was to assert that it turned William Gibson's fiction into fact.

Through claims that could not be fulfilled and laws that could not be considered constitutional, the Internet has emerged as a, if not *the*, new medium (the ~~mass~~ medium, which, because of its flexibility and variability put the term "mass" *sousrature*). These "false" claims did not simply mistake or propagate propaganda for reality; they affected Internet development and ideology, and these "virtualities" were (and still are) surprisingly resilient in the face of contradictory experiences. The same corporations selling the Internet as empowering sponsored debates on the digital divide; the lack of dot-com profitability did not affect belief in the new economy. The dot-coms turned into dot-bombs through a "fact," their unprofitability, that had always been known, and this fact challenges the idea that better accountability ensures better actions, that all we need is better information, more transparency. Indeed, as I discuss in detail in chapter 2, the notion that better information means better knowledge, which in turn means better action, founded and foiled the dream of the Internet as the ideal democratic public sphere.

The Internet, this book stresses, emerged as a medium (to end all mass media) through a particular stage of forces: the U.S. government's

---

41.   For examples, see George Landow, *Hypertext: The Convergence of Contemporary Critical Theory and Technology* (Baltimore, MD: Johns Hopkins University Press, 1992); The Smithsonian without Walls ⟨http://www.sliedu/revealingthings/⟩ (accessed May 1, 2004); Sherry Turkle, *Life on the Screen: Identity in the Age of the Internet* (New York: Simon and Schuster, 1995), and *Women and Performance* 17, ⟨http://www.echonyc.com/~women/Issue17⟩ (accessed June 8, 1999).

long-standing support of the Internet as a military and research network, and its decision in 1994 to privatize the backbone; the concurrent imagined and real expansion of technologies such as virtual reality (VR); the conflation of the Internet with cyberspace; a thriving personal computer and software industry, which was able to slash prices through outsourcing to Asia and Mexico; interest by various media companies and telecommunications companies in merging and expanding their markets (made possible through the Telecommunications Act of 1996); technological advances that made the Internet more image friendly (Web, image-oriented browsers); and extreme coverage in other mass media. All these forces, combined with these theory-come-true moments, turned a network cobbled together from remnants of military and educational networks into an electronic marketplace, a library, an "information super-highway," a freedom frontier. Through this combination, technology—seemingly forever condemned after the nuclear age—became good once more. Technology became once again the solution to political problems.

Control and Freedom: Power and Paranoia in the Age of Fiber Optics thus interrogates these forces and these theory-come-true moments not merely to debunk them as fraudulent (simply debunking them is as problematic as simply promoting them) but rather to understand their effects and the practices they engender. It investigates these moments in order to understand the linking of freedom and democracy to control, and the justification of this linking through technologically determinist explanations. This is not to say that technology has no force—its structures and language have a profound effect on our world and us. This is to say instead that technological solutions alone or in the main cannot solve political problems, and the costs of such attempts are too high: not only do such solutions fail but their implementation also generalizes paranoia.

Fiber-Optic Networks

Even though technology is not a simple cause, examining its structures and its emergence closely can help us understand our current situation, which is why this book concentrates on and takes inspiration from fiber-optic networks. Theoretically, fiber-optic networks work the fundamental paradox of light. In them, light is both wave and particle: lasers emit particle-like light, whereas the glass transports wavelike light.[42] Fiber-optic networks thus represent the theoretical necessity of us-ing, rather than resolving, paradoxes. Fiber-optic networks also work the divide between physical and virtual locations. They physically span the globe, buried within oceans or spanning office buildings' ceilings, while at the same time carrying the light necessary for these other spaces. Unlike "information," fiber optics emphasize the physical necessity of location and the explosion of virtual locations. Moreover, as Neal Stephenson argues in his "hacker tourist" quest to track the laying of the longest fiber-optic cable in 1997, fiber-optic cables reconfigure our understanding of the "real" world. Stephenson sees cable laying as an attempt to turn Mother Earth into a huge motherboard.[43] Fiber-optic networks also en-gage the infamous last-mile problem. The speed of the last mile basically determines the speed of the connection. The local and the global are not independent; speed depends on traffic, noise, and previous wiring.

The age of fiber optics delineates a specific time range and corporate phenomenon. Videophone and dotcom hype drove the deployment of fiber optics. Put into experimental use in the 1970s, fiber optics trans-formed the long-distance telecommunications industry. MCI entered the long-distance market by investing in single-mode fiber-optic cables, while AT&T was still experimenting with multimode cable.[44] At first, the hopes for the videophone drove the development of broadband, and the great expectations surrounding high-bandwidth real-time applications (expecta-tions propagated by articles such as Stephenson's) seemed to turn bust to boom. The Internet instigated the frenetic laying of fiber-optic cable in the mid- to late 1990s. Much of this cable remains unused, however, and this "dark fiber," which Geert Lovink takes up so eloquently in his book of the same title, combined with vacant fiber-optic factories in North Car-olina, reminds us of fiber optics' place within the larger economic system.

---

42.   For more on fiber optics, see Palais, *Fiber Optic Communications*.

43.   Neal Stephenson, "Mother Earth Mother Board," *Wired* 4, no. 12, ⟨http://www.wired.com/4.12/ffglass.html⟩ (accessed January 1, 1999).

44.   For more on the role of fiber optics in the deregulation of the telecommuni-cations industry, see Jeff Hecht, "Three Generations in Five Years (1975–1983)," in *City of Light*, 176–200.

The age of fiber optics is quickly being displaced by wireless technologies (which also preceded it). Wireless technologies open up the possibility of touch: of being constantly caressed or bombarded by the signals around us—signals that only some connectors can translate into a signal. Wireless technology's dominance in the South, where infrastructure costs are high, is also a result of geopolitics. Viewed by some as a case of technological "leapfrogging" (those poorer countries avoid the mistakes of more advanced countries by moving immediately toward more advanced technologies), this phenomenon leaves its frogs more vulnerable to both the effects of nuclear war and surveillance. Fiber optics replaced copper in key systems not only because of their speed but also because of their insensitivity to electromagnetic pulse (fiber-optic cables do not radiate energy). Because fiber-optic cables are also difficult to tap mechanically, and because they are usually buried, they offer a more secure and reliable form of communication than wireless or copper; the United States and the United Kingdom bombed Iraq in February 2001 when it tried to complete a Chinese-engineered fiber-optic network.

The other chapters of this book analyze in more detail the relation between fiber-optic networks and control-freedom. Unlike Foucault's investigation, this work focuses on the impact of sexual "freedom" rather than the historical processes that led us to the ironic belief that sexuality—with its attendant call to "tell everything"—could liberate us. The following chapters bring together what we can and cannot see, what is on, behind, and beyond the screen.

The Interlude draws out the uncanny similarities between Daniel Paul Schreber's paranoid hallucinations of 1903 and the high-speed networks of 2003. Schreber's system—a communications network, which confuses "pictured men" with real ones and consists of light rays and a "writing down system" that records everything—parallels our current fiber optic technologies. Rather than resting with this parallel, the interlude argues that this literalization and generalization of paranoia leads elsewhere. If Schreber's paranoia, as Santner argues in *My Own Private Germany: Daniel Paul Schreber's Secret History of Modernity*, arose from his realization that power is rotten at its core, that the disciplines sustain the liberties, ours blinds us to the transformation of discipline and liberty into control and freedom.

Chapter 1, "Why Cyberspace?" addresses the discontinuities between the Internet as Transmission Control Protocol/Internet Protocol (TCP/IP), the Internet as popularly conceived of as "cyberspace," and William Gibson's fictional "cyberspace." Arguing that the Internet has little to nothing in common with cybernetics or Gibson's fiction and that it is not spatial, this chapter contends that cyberspace's power stems from the ways it plays with notions of place and space. Cyberspace maps the Internet as a perfect frontier, as a heterotopia. Cyberspace has also enabled certain critical thinkers to theorize users as flâneurs. In order to operate, however, the Internet turns every spectator into a spectacle: users are more like gawkers—viewers who become spectacles through their actions—rather than flâneurs. Users are used as they use. Through an analysis of TCP/IP, this chapter argues that the public/private binary has been supplanted by open/closed. The increasing privatization of space and networks is responsible for this supplanting and poses the most significant challenge to liberal democracy today. More positively, this chapter argues that the Internet also establishes "touch" between users, and that this touch and our vulnerabilities lay the foundation for democratic action. This democratic potential, however, is placed constantly at risk through the conflation of control with freedom.

Chapter 2, "Screening Pornography," analyzes the "Great Internet Sex Panic of 1995," the U.S. Federal and Supreme Court decisions on the Communications Decency Act, and online pornography. It contends that the "discovery" of online pornography and the government's attempts to regulate it led to the dot-com craze of the late 1990s. Through cyberporn, the Internet became a marketplace (of ideas and commodities) in which "bad" contact stemmed from "bad" content rather than the Internet's context/structure. Through cyberporn, the pedophile and the computer-savvy child became hypervisible figures for anxiety over the jacked-in computer's breaching of the home. Electronic contact, however, cannot be divided into the "safe" and the "dangerous" based on content because the risk of exposure underlies all electronic exchanges. The conflict between Web page content and form, especially apparent within pornographic Web sites, exposes the fact that Hypertext Markup Language (HTML), HTTP, and javascripts—and not user mouse clicks—predominantly control interactivity. Drawing on the work of Claude Lefort and Thomas Keenan, this chapter argues that the Internet's democratic potential actually lies in these risky (nonvisible) encounters between self and other, where neither of these terms is necessarily human.

Chapter 3, "Scenes of Empowerment," asserts that in order to sell the Internet as a more democratic or "free" space, promoters conflated technological and racial empowerment. Analyzing MCI's "Anthem" commercial and United Nations documents on the digital divide, chapter 3 argues that a logic of "passing" lies at the heart of this conflation. The Internet, rather than enabling anonymity, supposedly allows users to pass as the fictional whole and complete subject of the bourgeois public sphere. This narrative of passing threatens to render invisible the practices of the very people of color from whom the desire to be free stems, and to transform the desire to be free from discrimination into the desire to be free from these very bodies. It has also led paradoxically to race's emergence as a pornographic category—one passes as the other by consuming its objects of desire. This chapter ends by considering work by the digital collective Mongrel, which refuses to commodify or erase race. The collective's work questions the effectiveness and desirability of passing, and pushes the democratic potential of the Internet.

Chapter 4, "Orienting the Future," contends that U.S. and Japanese cyberpunk make electronic spaces comprehensible and pleasurable through the Orientalizing—the exoticizing and eroticizing—of others and other spaces. Through close readings of William Gibson's *Neuromancer* and Mamoru Oshii's *Ghost in the Shell*, chapter 4 insists that the disembodied "user" construct relies on another disembodiment—namely, the reduction of the other to data. Cyberpunk's global vision—its force as a cognitive map—stems from its conflation of racial otherness with localness. This chapter does not simply dismiss cyberspace and electronic communications as inherently Orientalist but rather investigates the ways in which narratives of and on cyberspace seek to manage and engage interactivity, for high-tech Orientalism is not simply a mode of domination but a way of dealing with—of enjoying—perceived vulnerability.

Chapter 5, "Control and Freedom," concludes the book by clarifying control-freedom and linking it to the rise of a generalized paranoia. It revisits the commercials addressed in chapter 3 in order to expose the paranoia driving them, and then moves to a closer analysis of freedom-control through readings of face-recognition technology and Webcams. Against the current conflation of freedom with safety, chapter 5 agrees with Jean-Luc Nancy that freedom is something that cannot be controlled, that cannot be reduced to the free movement of a commodity within a marketplace. To do so is to destroy the very freedom one claims to be protecting. Rather than simply agreeing with Nancy, however, this chapter argues that his philosophical notion of freedom works by making oppression metaphoric. Lastly, it contends that the changing role of race exemplifies our experience of control-freedom as sexuality.

Control and Freedom: Power and Paranoia in the Age of Fiber Optics does not merely criticize the Internet, or users' freedom. To claim that users are an effect of software is not to claim that users, through their actions, have no effect. Everyone uses: some use as they are used by fiber-optic networks; some have no access to them and yet are still affected by them. The fact that using makes us vulnerable does not condemn the Internet, for what form of agency does not require risk? The problem lies not with our vulnerability but with the blind belief in and desire for invulnerability, for this belief and desire blind us to the ways in which we too are implicated, to the ways in which technology increasingly seems to leave no outsides. From our position of vulnerability, we must seize a freedom that always moves beyond our control, that carries with it no guarantees but rather constantly engenders decisions to be made and actions to be performed.

INTERLUDE

1903

Daniel Paul Schreber—former Senatsprasident *of the third chamber of the Supreme Court of Appeals in Dresden, failed National Liberal candidate in the Reichstag elections of 1884, and son of the famous Leipzig orthopedist and educator Moritz Schreber—publishes* Memoirs of My Nervous Illness, *which documents his schizophrenia, and explains how he must be transformed into a woman and impregnated to save the human race. Supposedly written to inform his wife of his personal experiences and religious ideas, Schreber's memoirs are published against her and his psychiatrists' wishes because he believes an "expert examination of my body and observation of my personal fate during my lifetime would be of value both for science and the knowledge of religious truths.... [I]n the face of such considerations, all personal issues must recede."[1] The interest of twentieth-century "experts" such as Sigmund Freud, Jacques Lacan, Gilles Deleuze, Felix Guattari, and Friedrich Kittler, among many others, has vindicated Schreber's decision, although not as he expected. None believe him to have been the redeemer of humankind, though none view him as simply delusional. Rather, they see his text as key to understanding subjectivity, modernity, and/or capitalism, and often cast it as prefiguring their own theories. This resonance, to a large extent enabled by Schreber's education and means, has made him the most influential psychotic to date.[2]*

---

1. Daniel Paul Schreber, *Memoirs of My Nervous Illness*, trans. Ida Macalpine and Richard A. Hunter (Cambridge: Harvard University Press, 1988), 3.

2. Schizophrenia at the turn of, and well into, the twentieth century was considered to be a lower-class disease.

Schreber's understanding of humanity and its relation to God coincides with the (then) emerging science of neurobiology. "The human body," Schreber explains, "is contained in the nerves of bodies," which retain the memory of received impressions. These nerves, which start out white, become blackened through the "sins" of the person to which they belong. On death, which Schreber informs us is really a state of unconsciousness, these nerves or "tested souls" are purified of individual consciousness, often by transmigrating to other planets.[3] God, in consonance with and contrast to humanity, comprises only the purest and whitest nerves or "rays," which are infinite and eternal. In order to create, God cuts off part of his nerves. He does not, however, diminish since he reaps these rays—and their accumulated memories—once the nerves have been purified. Due to his enlightened and enlightening communication system's time delay, God does not usually "interfere directly in the fate of peoples or individuals." This time delay also means that God only knows corpses and thus holds many erroneous beliefs about living human beings. Regardless, this pattern of nonintervention ensures limited human freedom and God's healthy existence, for "to draw close to living mankind was connected with dangers even for God Himself."[4] These dangers amount to God's enslavement—an enslavement based on his "addiction" to the excited nerves of human beings.[5]

According to Schreber, his first psychiatrist, Dr. Paul Emil Flechsig (a brain anatomist specializing in the localization of nervous illness), established such a destructive and compulsive real-time contact between God and Schreber.[6] Con-

---

3.   The souls undergoing purification are called "tested souls," rather than the more intuitive "untested souls," due to the oddities of the "basic language," which is a version of high German marked by many euphemisms.

4.   Schreber, *Memoirs*, 23.

5.   Thus God, in Schreber's world, is neither omnipotent nor omniscient but rather caught in a system whose supreme law is the "Order of the World." This order, however, can also be transgressed temporarily, if not permanently. Regardless, there is no absolute freedom, even for God.

6.   Flechsig first transgressed against the Order of the World by establishing direct contact with Schreber's nerves and making them speak against Schreber's will "from without incessantly and without any respite" (Schreber, *Memoirs*, 55). This unnatural situation, Schreber conjectures, is related to soul murder, which

sequently, God has become Schreber's persecutor: God, desperate to free himself, seeks to drive Schreber mad because mad souls hold no power of attraction, but his efforts will not succeed, for it is against the "Order of the World" to destroy a man's sanity (for some reason the Order of the World, which was circumvented in order to establish the connection, cannot be circumvented in this case). And so the more God tries to leave Schreber, the more excited Schreber's nerves become; the more excited they become, the greater the attraction. In order to reestablish the Order of the World, Schreber must be implanted with "voluptuous" nerves so that he can be "unmanned" and impregnated by the lower God. His offspring will then repopulate the world (Schreber oscillates between believing that the world is destroyed and the people he meets are imaginary, and acknowledging that the world still exists). Schreber must therefore enjoy himself constantly, constantly think voluptuous thoughts, so that God will remain attached. Remarkably, God does not realize that maintaining contact is in his best interest because he cannot learn from experience (God is not a "smart" program). And so God's "policy of vacillation" continues, and Schreber remains barren.

Rest is not possible for Schreber because God has introduced a system of compulsive thinking, which denies Schreber "man's natural right to think nothing." God's rays constantly examine him, constantly ask him, What are you thinking of now? Although Schreber at first refused to respond because this question is essentially unanswerable, the rays have circumvented his resistance through a system of "falsifying my thoughts."[7] Schreber's nerves compulsively finish key phrases.[8] As Schreber somewhat mysteriously explains, this torturous "compulsive thinking" is caused by God's confusion of nerve language with human language, stemming from his erroneous belief that a human being's mental capacities are extinct

---

like hypnosis, imprisons someone's will (9). After establishing such contact, Flechsig enlisted God (or at least half of him: God is split into a lower God Ariman and an upper God Orzmund) into his scheme by establishing direct contact between God and Schreber in order to destroy Schreber's sanity, so that Flechsig could possibly violate his body.

7.   Schreber, *Memoirs*, 56.

8.   He likens this situation to the role of a parent sitting as an audience member during a public school examination, during which the parent cannot help but answer the questions posed to his child.

if that person is not "speaking" nerve language (there are two kinds of language: normal human speech and nerve language; nerve language is like silent speech).[9] *(This erroneous belief in turn stems from God's inexperience with live humans.) Regardless, to convince God he is not insane and to give himself some rest, Schreber repeats poetry and prose.*

This constant examination also has a written component. In order to test Schreber's sanity and torment him further, God records everything Schreber says, and his phrases are later recalled and studied.[10] *God's erroneous belief that thoughts can be exhausted, that a complete archive of thoughts is possible, grounds this writing-down system. Yet, thoughts constantly provoke new ones, making such an archive impossible. Despite this, material from this writing-down system is played back to Schreber whenever he thinks nothing, and when Schreber repeats a thought, rays are sent down with the phrase, "We have already got this" (written down). Repetition "in a manner hard to describe" makes the rays "unreceptive to the power of attraction of such a thought."*[11] *This system of attraction constantly demands the new, and writing has enabled a type of memory or storage within a memoryless system, albeit one that cannot learn from the past, even as it checks for repetition.*

Schreber's body universally attracts and purifies. Tested souls plunge into his body singly and in groups (indeed, the one constant in Schreber's universe seems to be his irresistibility). Rather than being harmed by them and their poison, however, he miraculously purifies them. Until they lose their individual memories, they exist as "little men," roaming over his body and seeing through his eyes; he can also "picture" things in his mind for them to see.[12] *These tested souls, like all aspects of Schreber's mental illness, are racialized: God's upper and lower parts are divided into the Caucasian and the Semitic; a Catholic sect that seeks to make Germany Slavic plunges into Schreber's body; the basic language (spoken in heaven) is based on German because Germans are now God's chosen people;*

---

9.   Schreber, *Memoirs*, 54.

10.   Ibid., 125.

11.   Ibid., 128.

12.   These little men also seem to be projections; at one point, Schreber believes that the entire world has been destroyed and the real human beings he sees are merely little men miraculously pictured for him.

and before Schreber, only the "wandering Jew" had gone through the ordeal of gender transformation for divine impregnation.

This paranoid system—which enables both xenophobia and cross-racial identifications, records every communication, and forces communication against one's will in a language one is not necessarily conscious of, and in which memories (stored in the nerves) are viewed as the stuff of human beings, God (who is neither omniscient nor omnipotent) has perversely taken an interest in all living things, and rays of light serve as communicative fibers between all human beings—seemingly characterizes the ideal system of fiber-optic control. Schreber's intensified body, his constant striving for pleasure, and his freedom "this side of bureaucratization and human dignity" also seem to coincide with freedom within fiber-optic control.[13] *Whereas Schreber had to fight for freedom of movement, however, we are free to move as we please, as long as we take our medication.*

---

13.   Friedrich Kittler, *Discourse Networks 1800/1900*, trans. Michael Metteer, with Chris Callens, (Stanford, CA: Stanford University Press, 1990), 303.

| 1 |
WHY CYBERSPACE?

In order to emerge as a medium of freedom, the Internet became cyberspace in the mid-1990s. John Perry Barlow, in his infamous "A Declaration of the Independence of Cyberspace," declared cyberspace "the new home of the Mind," outside the sovereignty of "governments of the Industrial World."[1] U.S. Federal and Supreme Court decisions in 1996 and 1997 delineated the "Nature of Cyberspace," arguing that cyberspace was the resulting whole of decentralized, global communications as experienced by users.[2] All the major U.S. broadcast and cable news channels documented this strange world of cyberspace in which science fiction collided with reality, and invited their audiences to join them in cyberspace by sending them e-mail. Movies such as *The Net* revealed the dangers of living in cyberspace, as did the flood of articles in major print sources about cyberstalking (there were also more upbeat ones on cybersex and cyberdating). The prefix "cyber-" proliferated madly, signaling that electronic communications made strange—and even perhaps wondrous—everyday activities. Cyberspace, as a virtual nonplace, made the Internet so much more than a network of networks: it became a place in which things happened, in which users' actions separated from their bodies, and in

---

1.   John Perry Barlow, "A Declaration of the Independence of Cyberspace." ⟨http://www.salon1999.com/08/features/declaration.html⟩ (accessed May 1, 1999).

2.   See United States District Court for the Eastern District of Pennsylvania. *American Civil Liberties Union v. Reno* (Cir. A. No. 96–0963), ⟨http://www.eff.org/pub/Censorship/Exon_bill/HTML/960612_aclu_v_reno_decision.html⟩ (accessed May 21, 1998) and *Supreme Court Opinion (No. 96–511): Reno v. American Civil Liberties Union et al.*, ⟨http://www.ciec.org/SC_appeal/opinion.shtml⟩ (accessed September 19, 1997).

which local standards became impossible to determine. It thus freed users from their bodies and their locations.

This chapter examines the "weirdness" of cyberspace by looking at the ways in which it, as a heterotopia, plays with notions of place and space. It then discusses the congruities and tensions between cyberspace and Internet protocols in order to explore the ways in which the Intenet is public. According to those against the Internet's privatization and subsequent commercialization and mass use, the Internet was truly public when the U.S. government owned it.[3] During this idyllic period, commerce was forbidden and TCP/IP developed with little regard for "security," since the "community" of users was small and select. When the Internet went public by being privatized in 1994–1995, telecommunications and cable companies began building backbones (MCI/WorldCom was the majority owner of the Internet backbone in 2002). The Internet, then, as the Supreme Court argued, became a shopping mall—a privately owned, publicly accessible space—and the entrance of cable companies as Internet Service Providers (ISPs) profoundly altered the backbone's status, since these ISPs closed their cables to competing traffic. The disappearance of publicly owned, publicly accessible spaces (where publicly owned means state owned) and the concurrent emergence of publicly accessible, privately owned spaces has driven the transformation of public/private to open/closed.[4] This transformation poses the most significant challenge to democracy today.

Where's the Space?

Cyberspace is an odd name for a communications medium. Unlike *newspaper* (news + paper) or *film*, it does not comprise its content or its physi-

---

3.   This notion of state owned as public reveals important differences between eighteenth- and twentieth-century notions of public: according to Jürgen Habermas, the public emerged in opposition to both government and private interests; Immanuel Kant considered the government to be private.

4.   This transformation is especially clear in discussions of "public art." To most curators, public art is not art made or owned by the public but rather art that can be readily seen by the public—art on large television screens in Times Square or on the outside of San Francisco's Mosconi Center, or art located inside glass-enclosed private buildings that can be seen from the outside. Transparency becomes key.

cal materials. Unlike *movies*, derived from "moving pictures," it does not explain its form; unlike *cinema* (short for cinematograph: Greek *kinhma*, *kinhmato* [motion] + *graph* [written]), it does not highlight its physical machinery. Further, unlike *television* (tele + vision; vision from afar), cyberspace does not explain the type of vision it enables, and unlike *radio*, it does not reference its means of transmission (radiation). Although all these names—newspaper, film, movies, cinema, television, and radio—erase sites of production, cyberspace erases all reference to content, apparatus, process, or form, offering instead a metaphor and a mirage, for cyberspace is not spatial. Like telephone conversations and unlike face-to-face ones, electronic communications do not take place within a confined space; contrary to turn-of-the-century parlance, you do not meet someone in cyberspace. Not only are there at least two "originary" places (the sender's and the recipient's computer), data travels as discrete packets between locations and can be cached in a number of places. At best, a hypothetical route (paradoxically called a trace) of an interchange can be produced using packets with stepped Time To Live (TTL) settings.[5] Thus, if understood as the hardware and the protocols needed to connect users or more properly "their" machines, cyberspace is constantly changing and fundamentally unmappable. If understood as the higher-level scripting languages used to create Web pages, cyberspace is spaceless. There is, as Lev Manovich contends, "no space in cyberspace": HTML and Extensible Markup Language (XML) place objects against one another in an aggregate fashion, without creating a continuous or coherent perspectival space.[6]

Also, the notion of a cybernetic space—a space of, for, or defined by cybernetics—does not compute. Norbert Wiener coined *cybernetics* in

5. To produce such a map, the trace route tool sends out a series of packets with increasing TTL values, starting at one "hop." Whenever the packet "dies," the router at which the packet expired sends a message to the originating machine. Since packets can take different routes through the network, this is not entirely reliable. The fact that this is called a trace is itself fascinating and the basis of an investigation of the "waning" of deconstruction.

6. Lev Manovich, *The Language of New Media* (Cambridge: MIT Press, 2001), 253.

1949 to encompass "the entire field of control and communication theory, whether in the machine or in the animal."[7] During World War II, the MIT Radiation Laboratory, which was developing *servomechanisms*, combined control techniques developed for long-distance telephony amplifiers and aircraft guidance with Wiener's theory of stochastic process feedback. After World War II, more sensitive yet stable control mechanisms were developed, as well as more precise (computationally-based) ways of understanding and modeling control systems.[8] In calling this new field cybernetics, based on the Greek term *kybernete* (steersman, governor), Wiener effectively linked his mathematicization of negative feedback control to the ancient art of navigation and James Watt's fly governor, casting humans as negative-feedback control mechanisms: both humans and machines allegedly sample output and rework it into their input in order to respond to changes in output—they both make decisions. (Crucially, the "freedom" within the system—its unpredictability—makes possible, and requires, these decisions.) Control systems reduce a diverse array of mechanical, electrical, and electromechanical configurations—and human situations—into indistinguishable black box diagrams. They erase errors and bottlenecks caused by human inattentiveness and absentmindedness, which capitalist industrialism, as Jonathan Crary argues in *Suspensions of Perception: Attention Spectacle, and Modern Culture*, created as problems. (Perhaps "smart" control devices that can reprogram themselves will reveal the servility behind popular notions of freedom and autonomy, and their stupidity will reveal our own.) Cybernetics internalizes communications engineering and externalizes the central nervous system: animal internal mechanisms are control systems, and electromechanical devices nervous conduction systems.[9] Cyberneticians have focused on producing

---

7.   Norbert Wiener, *Cybernetics, or Control and Communications in the Animal and the Machine*, 2nd ed. (Cambridge: MIT Press, 1961), 11.

8.   Control systems are "classically" designed using root-locus and frequency-response design techniques and designed more "modernly" via state-space techniques, which use computers to solve previously unsolvable ordinary differential equations directly and offer a more complete internal description of the system.

9.   As Wiener writes, "We are beginning to see that such important elements as the neurones—the units of the nervous complex of our bodies—do their work

humanlike robots, robotic attachments to humans, and/or information-based "organisms." In its systems theory variant, cybernetics has tried to explain and reclassify natural large-scale phenomena like plant communities as ecosystems. Hence a space for cybernetics, especially a space restricted to the so-called information superhighway or computer-mediated communications, seems nonsensical.

Part of the peculiarity of cyberspace stems from its sci-fi origins. William Gibson coined the term cyberspace in 1982, eleven years before the National Center for Supercomputing Applications (NCSA) introduced Mosaic, the first graphics-based Web browser. Although other media such as photography and television had literary precursors—or more precisely, works labeled precursors after the fact—no other medium takes its name from a fictional text. Inspired by the early 1980s' Vancouver arcade scene, Gibson sat at his typewriter and outlined a three-dimensional chessboard/consensual visual hallucination called the Matrix or cyberspace, in which corporations exist as bright neon shapes, and console cowboys steal and manipulate data. In his *Neuromancer*, cyberspace is a "graphic representation of data abstracted from the banks of every computer in the human system."[10] Gibson's vision of cyberspace, however, has little to nothing in common with the Internet—other than a common 1990s' fan base. Unlike the Internet of 2006, Gibson's cyberspace (and Neal Stephenson's Metaverse) is *navigable* and its breadth is conceivable. Console cowboys in Gibson's Sprawl trilogy (*Neuromancer*, *Count Zero*, and *Mona Lisa Overdrive*) control their data path as they *travel through* cyberspace—they move from graphic representation to graphic representation as though playing a video game (Stephenson's *Snow Crash* contains

---

under much the same conditions as vacuum tubes, their relatively small power being supplied from outside by the body's circulation, and that the bookkeeping which is most essential to describe their function is not one of energy" (*Cybernetics*, 15). This paralleling of neurons and vacuum tubes was also key to the development (by John von Neumann among others) of modern programmable computers (for more on this, see John von Neumann, *First Draft of a Report on the EDVAC*, ⟨www.cs.colorado.edu/~zathras/csci3155/EDVAC_vonNeumann.pdf⟩ [accessed September 12, 2003]).

10.   William Gibson, *Neuromancer* (New York: Ace Books, 1984), 51.

a high-speed chase in the Metaverse). Most important, in Gibson's fiction, users can visualize cyberspace's size and scope, even if they cannot know its intimate details. Near the end of *Mona Lisa Overdrive*, a large new data sector attaching itself to cyberspace causes "a good three-quarters of humanity ... [to jack in and watch] the show."[11] Also, Gibson's cyberspace is mystical: at the end of *Neuromancer*, cyberspace moves from being a "graphic representation of data" to the product of two artificial intelligences merging (this event is referred to in subsequent books as "When It Changed"). In *Count Zero*, vodou loa-like entities inhabit cyberspace: cowboys now strike deals with these loas and are "ridden" by them. The series ends with Bobby (the cowboy initiate in *Count Zero*) and his lover (simstim superstar Angela Mitchell)—both dead—adventuring into an alternate Centauri version of the Matrix (contained in an aleph) in order to understand "When It Changed." Gibson's fiction also places the origins of cyberspace elsewhere: it is the product of "primitive arcade games" and "military experimentation with cranial jacks," and his descriptions of "jacking in" rely heavily on other phenomena, such as sex and narcotics.[12]

As I argue more fully in chapter 4, if cyberspace and the Internet have become conflated, it is due not to inherent similarities between them but rather a *desire* to position Gibson's fiction as both an origin of and an end to the Internet—a desire stemming from cyberspace's seductive "orientation," its seductive *navigability*. The fact that the term *cyberspace* is fading and that the term *Internet* now describes all interconnected networks (rather than those running TCP/IP), indicates changes in high-speed networks' social and technical significance (the demise of local bulletin board systems [BBSs], virtual reality, and utopianism).[13] But cyberspace's fading does not mean that it was erroneous or unimportant, for mainstream uses of the term *cyberspace* diffused the Internet's "openness" in order to produce a mythical user.

---

11.   William Gibson, *Mona Lisa Overdrive* (Toronto: Bantam Books, 1988), 245.

12.   Gibson, *Neuromancer*, 51.

13.   As Craig Hunt notes, an Internet now no longer simply refers to a network built on IP but extends to "any collection of physical networks" (*TCP/IP Network Administration*, 2nd ed. [New York: O'Reilly, 1997], 3).

The U.S. judiciary's Communications Decency Act (CDA) decisions' "Findings of Facts" section most decisively moved cyberspace from a sci-fi dream to a legitimate name for a communications medium. In it, the district judges delineated the differences between the various "areas" within cyberspace (e-mail, the World Wide Web, Internet Relay Chat, and so on). Although the "Nature of Cyberspace" section focuses solely on the Internet, the term *cyberspace* was presumably chosen over the term *Internet* because it could include configurations, such as local area networks and BBSs, that do not necessarily link to the Internet as well as future VR technologies. In the mid-1990s, the term *cyberspace* also referred to unnetworked materials available via CD-ROM. Cyberspace emphasized the importance of user experiences rather than network technologies—a point that Margaret Morse investigates in *Virtualities*. Even given this legitimation, cyberspace remains part science fiction, not only because the visions of Gibson's Matrix (and later Stephenson's Metaverse) have not yet been realized (and never will) but also because cyberspace mixes science and fiction. Cyberspace—as a hallucinatory space that is always in the process of becoming, but "where the future is destined to dwell"—was key to the selling of the Internet as an endless space for individualism and/or capitalism, as an endless freedom frontier. The fact that the Internet, or more broadly computer-mediated communications, was and still is called cyberspace is truly remarkable.[14] Trying to understand cyberspace as merging space and cybernetics and then condemning it for misrepresenting "reality" thus misses the point: namely, that cyberspace alters space, cybernetics, and reality.

Cyberspace Now

Fundamentally unmappable and unlocatable, cyberspace is a free space in which to space out about space and place, fact and fiction. Electronic spaces, or more properly electronic interfaces that portray networks as spaces (which erase hardware and physical locations), displace old

---

14.   See Electronic Frontier Foundation, "'Censorship: Internet Censorship Legislation and Regulation, 1998' Archive," ⟨http://www.eff.org/pub/Censorship/ Internet_censorship_bills/1998_bills/⟩ (accessed March 4, 1999).

assumptions about space and place.[15] Based on symbolic addresses that are already translations of four eight-bit numbers represented in base10, which are themselves translations of voltage differences, alphanumeric locators—implemented in an effort to make computer networks more user-friendly—simultaneously impose, obfuscate, and displace location, address, area, and coordination. The Internet, through its URLs, disengages name from location and location from geography while offering the virtue of location. URLs do not correspond to geography; connecting to "arizona.princeton.edu" accesses Princeton University's system of UNIX servers, and each server is named after an Arizonan city, such as Phoenix. Princeton is located in New Jersey, not Arizona, yet this naming system makes Princeton's system coherent to the human user (one connects to arizona and then moves to a specific "city" within the state). The original UNIX machine phoenix probably referred to the mythical bird, whose frequent deaths and rebirths make it an apt name for a server. Regardless, after Princeton's e-mail needs could not be met by one machine, the state-based system was introduced. This naming system also reveals the fundamental arbitrariness of geographic names (there is no inherent

---

15.   See Dave Healy, "Cyberspace and Place: The Internet as Middle Landscape on the Electronic Frontier," in *Internet Culture*, ed. David Porter (New York: Routledge, 1997), 55–68. In it, he argues that cyberspace is the "'middle landscape' between space (empty frontier) and place (civilization) that allows individuals to exercise their impulses for both separation and connectedness" (66). He sees us as the "heirs not only of the primitivist philosopher Daniel Boone, who 'fled into the wilderness before the advance of settlement,' but also the empire-building Boone, the 'standard bearer of civilization'" (66). Placing cyberspace as a middle landscape, however, assumes that the Internet is a landscape to begin with, overlooking the work needed to construct it as such. Again, rather than mediating between space and place, the Internet allows us to space out about the difference between space and place. For more on space and cyberspace, see Kathy Rae Huffman, "Video, Networks, and Architecture," in *Electronic Culture: Technology and Visual Representation*, ed. Timothy Druckrey (New York: Aperture, 1996), 200–207; Chris Chesher, "The Ontology of Digital Domains," in *Virtual Politics: Identity and Community in Cyberspace*, ed. David Holmes (London: Sage Publications, 1997), 79–93; and Mark Nunes, "What Space Is Cyberspace? The Internet and Virtuality," in *Virtual Politics*, ed. David Holmes, 163–178.

reason why Arizona should be called Arizona—only historical ones), and calls into question notions of place and space. It is not simply, then, that cyberspace is not spatial but also that cyberspace, in its attempt to map telecommunications, complicates the map it was once supposed to emulate.

Although space and place are often used interchangeably (one definition for place according to the *Oxford English Dictionary* is a "two- or three-dimensional space"), place designates a finite location, whereas space marks an interval. Place derives from the Latin *platea* (broad way), and space derives from the Latin *spatium* (interval or a period). Because of this, place has been tied to notions of civilization, and space to freedom, emptiness, and frontiers.[16] Dave Healy, quoting from Yi-Fu Tuan's analysis of the New World, claims "place is security, space is freedom: we are attached to one and long for the other."[17] In contrast, Michel de Certeau, while asserting that place designates stability or proper relations, argues that space is a practiced place—space is what we experience, rather than that for which we long. Place is *langue*, and space *parole*; place is the overarching structure, and space the actual articulation. For de Certeau, space destabilizes place by catching it "in the ambiguity of an actualization, transformed into a term dependent upon many different conventions, situated as the act of a present."[18] So rather than space being unrealizable freedom, it is how we negotiate place—it is how we *do* place. Differentiating between maps and tours, de Certeau asserts that space is an "intersection of mobile elements." Whereas maps once indicated the itineraries that made them possible, they are now scientific documents that "collate on the same plane heterogeneous places." Tours, on the other hand, are

16.   This impossible yet open and ever-expanding frontier underlies Michael Hardt's and Antonio Negri's empire thesis—a thesis that deliberately parallels imperial power networks with communications networks: as they claim, "Perhaps the fundamental characteristic of imperial sovereignty is that *its space is always open*" (*Empire* [Cambridge: Harvard University Press, 2000], 416). Their empire thesis resonates with the boom in late-twentieth-century fiber-optic networks.

17.   Healy, "Cyberspace and Place," 57.

18.   Michel de Certeau, *The Practice of Everyday Life*, trans. Steven Rendall (Berkeley: University of California Press, 1984), 117.

a "discursive series of operations."[19] Although not unrelated, maps and tours offer differing spatial experiences: the former is totalizing, and the latter is contingent. If space is a practiced place, however, cyberspace practices space, displacing place and space, maps and tours. Cyberspace others place and space.

Cyberspace loosens place, for place is no longer stable or proper. Places disappear and/or move rapidly; creators/managers of Web pages often move or erase Web pages with little regard for those who have bookmarked or linked to them, or for search engines that have indexed them. Remarkably, given this uncertainty, the cheapness of data storage, and the prevalence of what Manovich calls "database complex"—the irrational desire to store everything—many are caught short by disappearing Web sites they could easily have cached locally. The metaphoric use of place blinds us to the Web's fluidity. Place is also unstable because places and addresses are not indexical: depending on the stored cookie, amazon .com will produce significantly different Web sites. More notoriously, Domain Name System (DNS) poisoning/spoofing displaces place. Employed during the Second Gulf War to redirect those seeking Al Jazeera's English-language site to a page stating, "God Bless Our Troops," DNS poisoning, which requires no programming skills, corrupts the DNS servers that translate URLs into numerical IP addresses.[20]

Cyberspace similarly loosens space from tours, paradoxically through navigation. According to Manovich, "New media spaces are always spaces of navigation."[21] Navigable spaces may predate new media, but spaces that must be traversed in order to be experienced and understood, Manovich maintains, epitomize new media. New media spaces, however, are fundamentally unnavigable. Users may navigate and control software interfaces, but this control compensates for, if not screens, the lack of control they have over their data's path.

---

19.   Ibid., 121, 119.

20.   For more details, see SecuriTeam.com's step-by-step guide to DNA poisoning, ⟨http://www.securiteam.com/securitynews/Domain_Hijacking__A_step-by-step_guide.html⟩ (accessed September 1, 2004).

21.   Manovich, *Language*, 252.

Users do not navigate their packets. Constantly opened and reopened so that routers may know where to send them, packets follow inefficient and insecure paths—they are resent and "killed" so they do not bounce endlessly from router to router (hence the TTL setting discussed earlier). In Ethernet networks, packets regularly collide and follow burnt-in algorithms to resend at different intervals; network interface cards (NICs) constantly "listen" to each other and listen for systemwide broadcasts, such as printers announcing they are "up." There is a real danger that the network will be swamped and crashed by packets that have little to do with users' "affirmative actions." If users "source route" their packets (that is, determine the exact path of their packets), they still do not usually successfully navigate the network, for many hosts will reject such packets as security risks.

Even new media reduced to its interface, as navigable space, rewrites the relation between space and tours. Consider, for instance, the experience of "surfing" or "browsing" the Web in 2005. Both Netscape Navigator and Microsoft's Internet Explorer rely on navigational icons. Netscape (other than version 7) features a lighthouse and a nautical steering wheel, while Explorer features a spinning "e" in the shape of a globe (prior to version 5, the "e" was a globe). When browsing the Web through Netscape, you are at the helm of the ship, with Netscape providing your guiding light (according to Wiener, you are the control mechanism). Browsing with Internet Explorer, you span the globe from space, with Microsoft serving as your Global Positioning System. In either case, by typing in an address, or by clicking from location to location, you teleport rather than travel from one virtual location to another, and the backward and forward icons do not move backwards and forwards between contiguous locations. This teleporting means that we catch locations "in the ambiguity of an actualization, transformed into a term dependent on many different conventions," in a new manner.[22] Through our moves to "stop" pages before they are completely loaded, we catch certain locations and situate them as contingent acts of a present. By moving from URL to URL, we cut the scenery or space between fixed locations, while at the

---

22.   de Certeau, *Practice*, 117.

same time experiencing this "gap" as an often unbearable space of time, in which we decipher the page that emerges bit by bit on the screen. Indeed, representing telecommunications networks' use as spatial journeys makes bearable the time lapse inherent to "high-speed" networks that contrary to commercial propaganda and (Paul Virilio's) theoretical idealization, are never really instantaneous, are never really "real time."

Timing Space

The Internet is as much about time as it is space, and we need to emphasize time in order to open up questions of using, and accentuate the similarities and differences between the Internet and television.[23] The Internet, to emerge as new media, was sold as remedying the ills of television and enabling mass enlightenment by breaking down the barrier between speaking and listening. Whereas television, owned by powerful interests, turned citizens into receivers and perverted the marketplace of ideas, the jacked-in computer enabled individual—race-, gender-, and infirmity-free—citizens to publish once more as scholars before the literate world. Whereas television induced zoning out and passivity, the Internet demanded active participation. Whereas television offered a time-constricted standardized schedule, the Internet offered its content twenty-four-hours-a-day/seven-days-a-week (24/7), and one searched for sites, rather than read a guide. Whereas television offered information that disappeared on contact, the Internet erased the difference between viewing and storing information. The Internet made media content concrete, savable, and exchangeable. Whereas television is organized around time, the Internet is paradoxically organized around space and memory.[24]

---

23.   A careful comparison between the Internet and television also must analyze the *space* of television.

24.   For more on television as organized around time, see Mary Ann Doane, "Information, Crisis, and Catastrophe," in *Logics of Television: Essays in Cultural Criticism*, ed. Patricia Mellencamp (Bloomington: Indiana University Press, 1990), 222–239; and Jane Feuer, "The Concept of Live Television: Ontology as Ideology," in *Regarding Television: Critical Approaches—An Anthology*, ed. E. Ann Kaplan (Frederick, MD: University Publications of America, 1983), 12–22.

Yet the Internet does not "free" television viewers from "programming" by offering its content 24/7, nor is it spatially rather than temporally oriented. Web pages, as mentioned previously, are not always "there," not only because they are often taken down but also because a Web server can only manage a finite number of simultaneous hits. The 1999 "denial of service" attacks on sites such as CNN.com and yahoo.com revealed this vulnerability most forcefully; consequently, "flooding" a site became a criminal offense rather than a nuisance or an act of civil disobedience (used by the Electronic Disturbance Theatre in their protest of the military suppression of the Zapatistas).[25] Flooding, however, occurs unintentionally; for instance, it is almost impossible to access mla.org the day the Modern Languages Association posts its October Joblist. Although engineers are working to increase the number of simultaneous hits a server can manage, and to build software that can recognize and respond to denial of service attacks, this number remains a significant limitation. The server's memory and maintenance/backup are similarly important. As well, the end speed of a fiber-optic network depends on the "last mile," for global networks are always experienced locally: many local conditions, such as traffic and twisted-pair bottlenecks, and even broadcast error messages, play critical, mostly overlooked roles. Many modem users cannot go "where they want to today" because they cannot access high-bandwidth sites. Most digital subscriber lines (DSL) and cable users download quickly, but upload at a speed at par with modem users. Users' experiences of telecommunications networks are thus "singular" because their experience varies with their service and their hardware as well as with the time they log on.

Emphasizing time rather than space (or lack thereof) also exposes the Internet's emulation of the televisual event. The Internet proffers

25. The Electronic Disturbance Theater pioneered electronic civil disobedience with their Floodnet software, which automated the process of "electronic sitins" (simultaneously and repeatedly reloading a targeted Web site to temporarily block access to them). Floodnet was targeted against the then Mexican President Zedillo's Web site on April 10, 1998, and against the Clinton White House Web site on May 10, 1998. See Stefan Wray, "The Electronic Disturbance Threater and Electronic Civil Disobedience," ⟨http://www.thing.net/~rdom/ecd/EDTECD.html⟩ (accessed January 1, 2003).

real-time events such as live chat with celebrities at a fixed hour or nightly performances on Webcam sites, for live or time-constrained events in cyberspace create something newsworthy, something worth paying for, something that exceeds the supposed "thereness" of electronic texts. The 1996 event "24 Hours in Cyberspace" enabled one of cyberspace's first appearances on all three broadcast evening news shows. Live events such as daily Webcam performances also short-circuit questions of indexicality or authenticity, as they also do on television.[26] Even though live Webcam performances are easily forged, these performances are accepted as indexing something "outside" the doctored images that dominate the Web (hence, the obsessive repetition of the adjective *real* to describe almost all simultaneous interactions). Emphasizing time, rather than space, also highlights the similarities between commercial Web pages and commercial television. Both essentially sell advertisers a portion of the viewer's time, as well as a space within their page or program, and the effectiveness of these advertisements is unknown: pundits who hold forth on the ineffectiveness of Internet advertisements regularly overlook the effects of channel zapping.

Despite the differences between surfing and zapping, using categories like flow, live, and segmentation to analyze the Internet offers new theoretical possibilities (surfing does, after all, derive from channel surfing). In "Reload: Liveness, Mobility, and the Web," Tara McPherson pursues this line of inquiry by arguing that the Web interface offers its users "volitional mobility." Volitional mobility yokes together the feeling of "presentness" stemming from the live with the feeling of choice. Beyond Web-browser interfaces, we can use flow to expose implied paths within pages, the workings of search engines and packet paths. The flow of telnet sessions/chat rooms comprises the overwhelming and quickly disappearing experience of "text flooding," of real-time text scrolling madly down one's screen.[27] Focusing on flow also redirects critical efforts toward the major-

---

26.   For more on this, see Feuer, "Concept."

27.   To make real-time communications more user-friendly, client software increasingly obscures this phenomenon.

itarian, undertheorized position of the "lurker," who clicks and searches rather than posts. To understand the pertinence of flow to the Internet—indeed, to understand more generally the Internet as a public medium—however, we need to explore the differences between the Web and listservs, chat rooms, and e-mail at the level of user interface and application. All these differences and the importance of time is repressed by "cyberspace."

Othering Space

Early on, cyberspace's supposed openness and endlessness were key to imagining electronic networks as a terrestrial version of outer space. Constructed as an electronic frontier, cyberspace managed global fiber-optic networks by transforming nodes, wires, cables, and computers into an infinite enterprise/discovery zone. Like all explorations, charting cyberspace entailed uncovering what was always already there and declaring it new. It obscured already existing geographies and structures so that space became vacuous yet chartable, unknown yet populated and populatable. Like the New World and the frontier, settlers claimed this "new" space and declared themselves its citizens—conveniently, there were no real natives (just virtual ones, created by cyberpunk).[28] Advocacy groups, such as the Electronic Frontier Foundation, used the metaphor of the frontier to argue that cyberspace lay both outside and inside the United States, since the frontier effectively lies outside government regulation yet also within U.S. cultural and historical narratives. Moreover, cyberspace as a terrestrial yet ephemeral outer space turned attention away from national and local fiber-optic networks already in place toward dreams of global connectivity and postcitizenship. Those interested in "wiring the world" reproduced—and still reproduce—narratives of "darkest Africa" and civilizing missions. These benevolent missions, aimed at alleviating the disparity between connected and unconnected areas, covertly, if not overtly, conflate spreading

---

28.   For settlers' claims, see Barlow, "A Declaration of the Independence of Cyberspace"; Cleo Odzer, *Virtual Spaces: Sex and the CyberCitizen* (New York: Berkley Books, 1997); and Howard Rheingold, *The Virtual Community: Homesteading on the Electronic Frontier* (Reading, MA: Addison-Wesley, 1993).

the light with making a profit.[29] Through this renaming, cyberspace both remaps the world and makes it ripe for exploration once more.

Cyberspace as a frontier others space.[30] According to Foucault, in "Of Other Spaces," other spaces are heterotopias: they are "counter-sites, a kind of effectively enacted utopia in which the real sites, all the other real sites that can be found within the culture, are simultaneously represented, contested, and inverted. Places of this kind are outside of all places, even though it may be possible to indicate their location in reality."[31] Cyberspace lies outside all places and cannot be located, yet it exists. One can point to documents and conversations that "take place" in cyberspace, even if cyberspace makes such phrases catachrestic.[32] Moreover, cyberspace as a heterotopia simultaneously represents, contests, and inverts public spaces and places. On the one hand, as Manovich argues, the lack

---

29.　For *Wired*'s version of the civilizing mission, see Jeff Greenwald, "Wiring Africa," ⟨http://www.wired.com/wired/archive/2.06/africa.html⟩ (accessed May 1, 1999); John Perry Barlow, "Africa Rising," ⟨http://www.wired.com/wired/archive/6.01/barlow.html⟩ (accessed May 1, 1999); Nicolas Negroponte, "The Third Shall Be First," ⟨http://www.wired.com/wired/archive/6.01/negroponte.html⟩ (accessed May 1, 1999); and Neal Stephenson, "Mother Earth Mother Board," *Wired* 4, no. 12, ⟨http://www.wired.com/4.12/ffglass.html⟩ (accessed January 1, 1999).

30.　In defining cyberspace as a heterotopia instead of a utopia, I am responding to critics of the Internet such as James Brook, Iain Boal, and Kevin Robins who insist that the Internet is not a utopia, and that the mythology of the Internet must be debunked/demystified. Whereas they seek to put "sociology before mythology" and look at cyberspace's relation to the "real world," I argue that its mythology is precisely what links it to the real world, not as a regression or fantasy but rather as a public space. This is not to say that sociology is unimportant. This is to say that it must not be an either/or but both at once.

31.　Michel Foucault, "Of Other Spaces," trans. A. M. Sheridan Smith, *diacritics* (Spring 1986): 24.

32.　Drawing from Foucault's introduction to *The Order of Things*, Diana Saco similarly argues that cyberspace is a heterotopia in *Cybering Democracy: Public Space and the Internet* (Minneapolis: University of Minnesota Press, 2002). Specifically, she views cyberspace as "an in-between space of contradiction and contestation: one that mimics or simulates lived spaces but that calls those inhabited spaces into question" (76).

of space in cyberspace reflects the general U.S. apathy toward communal or public spaces:

The spatialized Web envisioned by VRML (itself a product of California) reflects the treatment of space in American culture generally, in its lack of attention to any zone not functionally used. The marginal areas that exist between privately owned houses, businesses and parks are left to decay. The VRML universe, as defined by software standards and the default settings of software tools, pushes this tendency to the limit: it does not contain space as such but only objects that belong to different individuals.[33]

On the other hand, there are marginal nondecaying virtual locations. "Decrepit places" are not marginal nonspaces in between privately owned pages but rather pages that are no longer updated, that list last year's lectures as next year's coming events: pages that no one seems to own. The visibility and portability of HTML code also troubles possession. Given that one can easily copy and replicate another's "private object," especially since one downloads what one views, the "ownership" of virtual items is not easy to define; possession is not exclusive (yet).

Cyberspace also alters heterotopias because its mirroring function is not indexical. According to Foucault:

I believe that between utopias and these other sites, these heterotopias, there might be a sort of mixed, joint experience, which would be the mirror. The mirror is, after all, a utopia, since it is a placeless place. In the mirror, I see myself there where I am not, in an unreal, virtual space that opens up behind the surface; I am over there, there where I am not, a sort of shadow that gives my own visibility to myself, that enables me to see myself there where I am absent: such is the utopia of the mirror. But it is also a heterotopia in so far as the mirror does exist in reality, where it exerts a sort of counteraction on the position that I occupy. From the standpoint of the mirror I discover my absence from the place where I am since I see myself over there. Starting from this gaze that is, as it were, directed toward me, from the ground of this virtual space that is on the other side of the glass, I come back toward myself;

---

33.   Manovich, *Language*, 258.

I begin again to direct my eyes toward myself and to reconstitute myself there where I am. The mirror functions as a heterotopia in this respect: it makes this place that I occupy at the moment when I look at myself in the glass at once absolutely real, connected to all the space that surrounds it, and absolutely unreal, since in order to be perceived it has to pass through this virtual point which is over there.[34]

Cyberspace functions as a utopia because it enables one to see oneself—or at the very least, one's words or representations—where one is not. In a "live chat," conversations take place in a virtual chat room. On a "home page," imaginary or representative images of oneself or one's possessions make themselves at home. Cyberspace functions as a heterotopia or countersite because it, or more precisely its interface, actually exists; but rather than making one's actual place both connected and unreal, cyberspace absents oneself from one's actual physical location: when one is on a MOO such as LambdaMoo, one is supposedly in a living room, hot tub, sex room, or nightclub. Howard Rheingold, explaining virtual community, declares, "We do everything people do when people get together, but we do it with words on computer screens, leaving our bodies behind."[35] This disappearing body supposedly enables infinite self-re-creation and/or disengagement, and poses the question, Where am I really?[36] If I am single-

---

34.   Foucault, "Other Spaces," 24.

35.   Howard Rheingold, "A Slice of Life in My Virtual Community," in *Big Dummies' Guide to the Internet: A Round Trip through Global Networks, Life in Cyberspace, and Everything*, textinfo edition 1.02 (September 1993), ⟨http://www.hcc.hawaii.edu/bdgtti/bdgtti-1.02_18.html#SEC191⟩ (accessed June 1, 1999).

36.   For more on the question of Where am I really? see Sherry Turkle, *Life on the Screen: Identity in the Age of the Internet* (New York: Simon and Schuster, 1995), and Allucquère Rosanne Stone, *The War of Desire and Technology at the Close of the Mechanical Age* (Cambridge: MIT Press, 1995). Many critics have also questioned the notion of the disappearing or virtual body. For instance, Vivian Sobchack concentrates on the ways in which pain reminds us that we are not simply virtual bodies (see "Beating the Meat/Surviving the Text, or How to Get out of This Century Alive," in *Cyberspace/Cyberbodies/Cyberpunk: Cultures of Technological Embodiment*, ed. Michael Featherstone and Roger Burrows [London: Sage Publications, 1995], 205–214). For more on the virtual/nonvirtual body, see

mindedly participating in an online conversation, am I not absent from my physical location? Do I shuttle between various "windows"—real life and VR? Further, if I am a female paraplegic online, but a male psychiatrist off-line, who am I really? These questions have led Sherry Turkle and Sandy Stone to theorize online interactions as normalizing or disseminating multiple personality disorder, as "shattering" real and mirror images. Stone argues that online interactions break the state's "warranting" between subject and body; Turkle contends that the experience of online persona concretizes postmodernism and can be therapeutically beneficial. Although Turkle's and Stone's work have been key to the study of MUDs and other online environments, linked to the early promise of VR, their foci overlooks the ways in which these images become spectacles, or circulating texts—not linked to our personalities or subjectivities—in their own right. Cyberspace does more than reflect back; it is more than a virtual location we traverse in order to reconstitute ourselves.

In the early to mid-1990s, cyberspace was marked as a heterotopia of compensation—as a space for economic, social, or sexual redress that simultaneously represented, contested, and inverted all other "real" spaces.[37] According to Turkle, young adults on MUDs often built virtual

---

Balsamo, "Forms of Technological Embodiment: Reading the Body in Contemporary Culture," in *Cyberspace/Cyberbodies/Cyberpunk*: *Cultures of Technological Embodiment*, eds. Michael Featherstone and Roger Burrows, 215–237; Michelle Kendrick, "Cyberspace and the Technological Real," in *Virtual Realities and Their Discontents*, ed. Robert Markley (Baltimore, MD: Johns Hopkins University Press, 1996), 143–160; Katie Argyle and Rob Shields, "Is There a Body on the Net?" in *Cultures of Internet: Virtual Spaces, Real Histories, Living Bodies*, ed. Rob Shields (London: Sage Publications, 1996), 58–69; Theresa M. Senft, "Introduction: Performing the Digital Body—A Ghost Story," *Women and Performance* 17, ⟨http://www .echonyc.com/~women/Issue17/introduction .htm⟩ (accessed June 8, 1999); and the articles collected in Holmes, ed., *Virtual Politics*: *Identity and Community in Cyberspace*, "Part I: Self, Identity, and Body in the Age of the Virtual."

37. Foucault in "Other Spaces" divides heterotopias into crisis heterotopias (boarding schools and honeymoons), heterotopias of deviance (rest homes and prisons), heterotopias of illusion (nineteenth-century brothels), and most important for my purposes, heterotopias of compensation (colonies).

representations of economic rewards denied them in real life. In the early 1990s, a college education did not guarantee well a paying job, but "MUDs [got them] back to the middle class" (*The Sims* seems to perform the same function in the 2000s).[38] Cyberspace enables virtual passing, allowing us to compensate for our own (perceived) limitations by passing as others online—and this online passing consequently affects how we imagine our real bodies.[39] Although MUDs have faded with the rise of the Web, instant messaging, and graphical simulations, the structure of virtual passing, through chat rooms/instant messaging, graphical worlds, or blogs, remains in place. This virtual passing promises—against the grain of technology—to protect our "real" bodies and selves from the glare of publicity. If those "in the public eye" have had to trade their privacy for public exposure, if they have spread their images at the risk of reducing their existence to proliferating images, cyberspace, by denying indexicality, seems to enable unscathed participation. Such passing moves one to a simpler, arbitrary, and less encumbered space in which one's representation and actuality need not coincide. Passing permits an imitation indistinguishable from the "real thing," yet completely separate from it—one passes when one's inner and outer identities cannot or do not coincide, or when one does not want them to coincide, thus also *creating* the notion of an inner and outer self. Importantly, passing is a form of agency, which brings together the two disparate meanings of agency: the power to act, and the power to act on another's behalf. In doing so, it reveals the

---

38.   Turkle, *Life*, 240.

39.   There is a rich body of work on nonvirtual passing; see, in particular, Adrian Piper, "Passing for White, Passing for Black," *Transition* 58 (1993): 4–32; Amy Robinson, "It Takes One to Know One: Passing and Communities of Common Interest," *Critical Inquiry* 20, no. 4 (1994): 715–736; Judith Butler, "Passing, Queering: Nella Larsen's Psychoanalytic Challenge," in *Bodies That Matter: On the Discursive Limits of "Sex"* (New York: Routledge, 1993), 167–185; Samira Kawash, "The Epistemology of Race: Knowledge, Visibility, and Passing," in *Dislocating the Color Line: Identity, Hybridity, and Singularity in African-American Literature* (Stanford, CA: Stanford University Press, 1997), 124–166; and the essays collected in Elaine K. Ginsburg, ed., *"Passing" and the Fictions of Identity* (Durham, NC: Duke University Press, 1996).

tension inherent to agency, the ways in which it is compromised even when it is effective, the ways in which agency is most forceful when mediated.

Cyberspace as a heterotopia of compensation follows Foucault's description (if not the reality) of other compensatory spaces, such as the colonies. Drawing on Puritan societies in New England and the Jesuits of Paraguay, Foucault argues that compensatory heterotopias represent a space of pure order. They are "as perfect, as meticulous, as well arranged as ours is messy, ill constructed, and jumbled." They are "absolutely perfect other spaces." Foucault depicts the Jesuit colonies in South America as "marvelous, absolutely regulated colonies in which human perfection was effectively achieved … in which existence was regulated at every turn."[40] The Web's transformation into an e-commerce paradise exemplifies the portrayal of the Internet as an absolutely perfect other space. Online, there are no crowds or obnoxious salespeople, there are no parking lots or mall corridors to negotiate. Also, unlike a store, everything is displayed; everything is findable, searchable, and orderable. Search engines make the Internet seem a perfect archive.

Foucault, however, glosses over the fact that this placing of pure order simultaneously obfuscates—if not annihilates—other spaces/places already in place, such as Native America (in general, the subordination and erasure of Native America grounds notions of the open frontier). At a fundamental level, cyberspace emerges through the erasure not only of hardware differences, but also disorderly hardware itself. In terms of hardware, the Internet does not exist—or, to be more precise, the difference the Internet makes cannot be recognized; at this level, machines influence each other through electromagnetic interference (in terms of EMI, a refrigerator and a computer are both communications devices), power (voltage*current) and resistance (voltage/current) are technical terms, and resistance is necessary for circuits to operate (as resistance goes to zero, current goes to infinity). Even at the cleaner level of logic gates, which erases circuit particularities, the difference the Internet makes makes no difference—a gate is a gate is a gate. Importantly, these other spaces do

---

40. Foucault, "Other Spaces," 27.

not completely dissolve but rather continually threaten "pure order." Because Puritan societies had to defend themselves against indigenous populations, their utopia was never effectively realized.[41] Voltage differences are imprecise and continuous (von Neumann himself argued that computers are *both* analog and digital); alternating currents generate EMI, and thus the possibility of crosstalk; plugging boards into sockets causes potentially damaging voltage spikes.[42] Regardless—and perhaps because of the difficulty of maintaining heterotopias—Foucault describes the boat as "the heterotopia *par excellence*." The boat is exemplary because it is "a floating piece of space, a place without a place, that exists by itself, that is closed in on itself and at the same time is given over to the infinity of the sea and, from port to port, from tack to tack, from brothel to brothel, it goes as far as the colonies in search of the most precious treasures they conceal in their gardens."[43] Foucault's privileging of the boat and nautical navigation resonates with Wiener's privileging of kybernete or governors, but also Jean-Luc Nancy's description of freedom as an experience that resonates with piracy. According to Nancy,

Experience is an attempt executed without reserve, given over to the *peril* of its own lack of foundation and security in this "object" of which it is not the subject but instead the passion, exposed like the pirate (*peirātēs*) who freely tries his luck on the high seas. In a sense, which here might be the first and last sense, freedom, to the extent that it is the thing itself of thinking, cannot be appropriated, but only "pirated"; its "seizure" will always be illegitimate.[44]

---

41.   As Hardt and Negri similarly point out, the U.S. view of empire as ever expanding depends on the deliberate and brutal ignorance of Native America (*Empire*, 169–170).

42.   John von Neumann, *Papers of John von Neumann on Computing and Computer Theory*, eds. William Aspray and Arthur Burks (Cambridge, MA: MIT Press, 1987), 400.

43.   Foucault, "Other Spaces," 27.

44.   Jean-Luc Nancy, *The Experience of Freedom*, trans. Bridget McDonald (Stanford, CA: Stanford University Press, 1993), 20.

Most important for Foucault, "In civilizations without boats, dreams dry up, espionage takes the place of adventure, and the police take the place of pirates."[45]

Cyberspace, then, offers wet dreams of exploration and piracy—the lure of freedom, but also its risks. It indicates the possibility of radical change—of refiguring the world so that it is perhaps not perfect but more livable.[46] Rather than freedom as experience, however, the wet dreams of exploration and piracy most often reduce to freedom as capitalist control. According to David Brande, cyberspace, through its limitless opportunity and open spaces, reinvigorates capitalism.[47] Cyberspace ends the narratives of the end; it ends narratives of postmodern/postindustrial society's ennui and exhaustion. It proffers direction and orientation in a world disoriented by technological and political change, disoriented by increasing surveillance and mediation. Yet cyberspace also disseminates what it would eradicate; it reflects back what it would deny. Cyberspace perpetuates the differences and contingencies it seeks to render accidental. Passing in cyberspace does not adequately protect viewers from becoming spectacles, from being in public. Instead, in order to maintain the fiction of the all-powerful user who *uses*, rather than *is used by*, the system, narratives on and about cyberspace focus the user's gaze away from its own vulnerability and toward others as spectacle.

Gawkers

The user, popularly understood as couch-potato-turned-anonymous-superagent, requires much online and off-line intervention. Browsers

---

45. Foucault, "Other Spaces," 27. Boats, of course, also have an alternate history that place them as dystopian heterotopias. The Middle Passage reveals the dreams enabled by boats as nightmares. Neal Stephenson plays on both images of boats in *Snow Crash* (New York: Bantam Books, 1992).

46. As mentioned in the introduction, this notion of reconstructing the world so it is more livable resonates with Freud's diagnosis of paranoid reconstruction in his case study of Daniel Paul Schreber.

47. David Brande, "The Business of Cyberpunk: Symbolic Economy and Ideology in William Gibson," in *Virtual Realities and Their Discontents*, ed. Robert Markley, 100–102.

deliberately conceal the constant exchange between so-called clients and hosts through spinning globes and lighthouses at which we gaze as time goes by. Recent critical theorizations of the user as a flâneur or an explorer similarly sustain the fiction of users as spectators rather than spectacles, or at least involuntary producers of information. Manovich, for instance, argues that flâneurs and explorers are the two major phenotypes for the user. Game users, navigating virtually empty yet adventure-filled spaces, mimic explorers in James Fenimore Cooper's and Mark Twain's fictions. The Web surfer is Charles-Pierre Baudelaire's flâneur: a perfect spectator, who feels at home only while moving among crowds. The reincarnated flâneur/data dandy, notes Manovich, "finds peace in the knowledge that she can slide over endless fields of data locating any morsel of information with the click of a button, zooming through file systems and networks, comforted by data manipulation operations at her control."[48] Although this trajectory is important, it is also important (as Manovich himself contends) to explore its limitations, especially since *flâneurie* depended in large part on Cooper's portrayal of the frontier and since a flâneur would never post.

To be a perfect spectator, one must see, but not be seen, "read" others and uncover their traces, but leave none of one's own. As Baudelaire argues,

For the perfect flâneur, . . . it is an immense joy to set up house in the heart of the multitude, amid the ebb and flow. . . . To be away from home, yet to feel oneself everywhere at home; to see the world, to be at the center of the world, yet to remain hidden from the world—such are a few of the slightest pleasures of those independent, passionate, impartial [!] natures which the tongue can but clumsily define. The spectator is a *prince* who everywhere rejoices in his incognito.[49]

48.   Manovich, *Language*, 274.

49.   Charles-Pierre Bandelaire, quoted in Walter Benjamin, *The Arcades Project*, trans. Howard Eilan and Kevin McLaughlin (Cambridge: Harvard University Press, 1999), 443.

As an impartial, unobserved observer, the flâneur asserts his independence from the urban scenes he witnesses. As Tom Gunning maintains, "The flâneur flaunt[s] a characteristic detachment which depend[s] on the leisurely pace of the stroll and the stroller's possession of a fund of knowledge about the city and its inhabitants."[50] Flâneurs unobtrusively walk turtles on leashes. On the so-called information superhighway, flâneurs and turtles would both be roadkill (metaphorically of course and importantly), for the Internet makes it impossible to be "at the center of the world" yet remain hidden (if it was ever possible elsewhere). Online, everyone automatically produces traces; every search produces a return address.[51] (Arguably, the increasing prevalence of surveillance cameras and satellite imaging makes urban *flâneurie* impossible.) As well, Manovich's "flâneur," who zooms from place to place, manipulating data with ease, follows in the footsteps of U.S. hard-boiled detectives, rather than nineteenth-century flâneurs. Regardless, the gawker, rather than the flâneur or the detective, is the more compelling model for users.[52]

The gawker (the *badaud*, which Howard Eilan and Kevin McLaughlin translate as rubberneck in Walter Benjamin's *The Arcades Project*), unlike the flâneur, is not independent. Quoting Victor Fournel's description of the badaud, Benjamin writes, "The average flâneur is always in full possession of his individuality, while that of the rubberneck disappears, absorbed by the external world, ... which moves him to the point of intoxication and ecstasy. Under the influence of the spectacle, the rubberneck becomes an impersonal being. He is no longer a man—he is the public; he is the crowd."[53] The gawker, captured by commodities, stands and stares. Like

---

50. Tom Gunning, "From Kaleidoscope to the X-Ray: Urban Spectatorship, Poe, Benjamin, and Traffic in Souls (1913)," *Wide Angle* 19, no. 4 (1997): 25–63.

51. An anonymizer will allow users to cover their tracks by erasing all record of interaction between an individual client and itself. Servers that track users therefore only know that the anonymizer site has contacted them. Installing a Trojan horse program on the client's computer can circumvent this erasure. Also, the user must trust that the anonymizer site is actually erasing everything.

52. In making this argument, I am drawing from Gunning's analysis of early filmgoers in "From the Kaleidoscope to the X-Ray."

53. Benjamin, *Arcades*, 429.

the lurker, the gawker is inundated by, and part of, the ongoing flood of information; like the lurker, the gawker is the object of someone else's gaze—treated as part of the crowd. The myth of superagent users, who dismantle and engage the code, who are explorers rather than explored, screens the lurker's vulnerable position. This myth tries to convince the user-cum-lurker that it is a flâneur, who leaves no traces as it observes, or when not "lurking," it is the detective, the active searcher of information. This myth both emphasizes user control and fosters paranoia, for if the user can go anywhere it wants, cannot someone else with more knowledge and skill track the user?

In order to circumvent this paranoid doubt, or any admission of vulnerability, Internet promoters produce spectacular spectacles, or at the very least sites that emphasize the agency of the user and not the server. Literary representations of cyberspace and early Net theory similarly promised the spectacular. Imagining "new" encounters between computer and humans, human and humans, cyberpunk literature, which originated the *desire for* cyberspace, if not cyberspace itself, seductively denied representation through dreams of disembodiment.[54] Cyberpunk offers unnerving yet ultimately readable "savage" "otherness" in order to create the mythic user. These narratives *romanticize* networks along with gritty city streets and their *colorful* inhabitants. In them, badass heroines and geek-cool hackers navigate through disorienting urban and virtual-as-urban landscapes, populated by noble and not-so-noble savages. Rather than a happy future, cyberpunk's future is edgy and vaguely dystopian. Rather than happy consensus-driven spaces in which differences disappear, cyberpunk spaces are pockmarked by racial and cultural differences that may be vaguely terrifying, but are ultimately readable and negotiable. Rather than brushing aside fear of strange locations, strangers, and their dark secrets by insisting that we are all the same, they, like the detective fiction on which they are often based, make readable, trackable, and solvable the lawlessness and cultural differences that supposedly breeds in

---

54. This idea of something originating the desire for, rather than the thing itself is drawn from Geoffrey Batchen's groundbreaking and insightful analysis of the "origins" of photography, *Burning with Desire: The Conception of Photography* (Cambridge: MIT Press, 1997).

crowds and cities. Cyberpunk hero/ines seek to classify and navigate through landscapes by reducing others to their markers of difference. In the end, these spaces, for all their unfamiliarity, are reduced to humanly accessible information—to a vast virtual library. Indeed, racial and ethnic differences, emptied of any link to discrimination or exclusion, make these spaces "navigable" yet foreign, readable yet cryptic. Difference as a simple database category grounds cyberspace as a "navigable space"; through racial difference we steer, and sometimes conquer. This navigate-by-difference narrative, expanded on in chapter 4, is high-tech Orientalism. High-tech Orientalism is an agoraphobic response to public, disorienting networks.

If the Internet is still public—that is, an indeterminate space that belongs to no one—it is because the Internet is a protocol, is TCP/IP. TCP/IP combines the names of two specific protocols, but refers more generally to an open four-layer protocol that transmits data between differing networks (Ethernet, ATM, and so forth) without the end computers determining the routes between them (see figure 1.1). Although

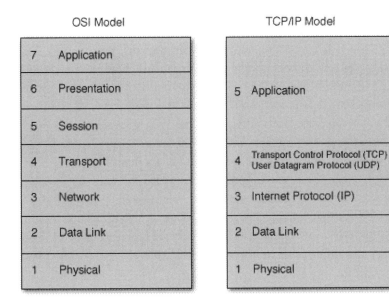

| Figure 1.1 |
TCP/IP architecture, ⟨http://www.firewall-software.com/firewall_faqs/firewall_network_models.html⟩

Arpanet did not officially require TCP/IP until 1983, well after its debut in 1969, TCP/IP has become the technical linchpin of global telecommunications networks, enabling networks running on different physical media (thick and thin coax cable, fiber, twisted pair, radio, satellite, and so on) as well as running different networking standards (Ethernet, token ring, and so on) to communicate. TCP/IP does this by creating a series of "envelopes" around the data it sends and through a series of procedures it puts in place to confirm reception. Each layer of this system adds a header (which contains source and destination information) to its payload (which includes user-generated data, if there is any, plus headers from the other layers). The bottom network access layer, which TCP/IP usually obscures, translates between hardware Media Access Control (MAC) addresses and IP addresses, and breaks up packets if they exceed the physical limitations of a specific network. IP addresses are not permanently assigned to machines. Thus, in order for the packet to be routed at the most basic level, tables must translate between IP addresses and MAC addresses, which are—or at least are supposed to be—singular. Because the bottom layer deals with MAC addresses, its source and destination machines must be on the same network, which means this header will change frequently in a packet's "lifetime." The bottom layer's payload contains the header and payload of the next layer, IP.

IP, which is connectionless and unreliable, packages network traffic into IP datagrams and defines the rules for moving them through the network.[55] Its header includes the source and destination IP address as well as information regarding the protocol of the next layer, its TTL, and many other variables. IP's routing strategies vary from simple static tables to external gateway protocols, such as Border Gateway Protocol, which can base routing decisions on security concerns or political and/or economic alliances (the system of policy-blind "core gateways" is long gone; many ISPs now have private agreements for data transport). The next layer, the transport layer, can either follow User Datagram Protocol (UDP) or TCP. These protocols operate at the host-to-host level, ensuring that the information transported by IP interacts correctly with the

---

55. See Pete Loshin, *TCP/IP Clearly Explained*, 3rd ed. (San Diego, CA: Academic Press, 1999), 126.

application layer. TCP is far more reliable than UDP, for it creates a "virtual circuit" between the source and destination machines. This virtual circuit enables IP tunneling, in which TCP/IP networks via HTTP carry encrypted information between two hosts. IP tunneling also enables corporations to develop Virtual Private Networks (VPNs) and thus, according to Saskia Sassen, turn the public network into a private connection.[56] Regardless of these private infringements, TCP/IP is still essentially open, but this openness—or to be more precise, readability—is constructed. All protocols are open to the extent they are protocols rather than proprietary standards, but TCP/IP's extreme readability stems from an early decision to handle hardware issues at lower levels and application-specific security issues, like encryption, at higher ones. Indeed, it is at the highest layer— the application layer—that more intrusive control, from cookies to spyware, is enabled.

Importantly, technological protocols are not ultimate limits: they change; they are circumvented. There is no fundamental nature of the Internet. The adage "the Net interprets censorship as damage and routes around it," and Mark Godwin's assertion that the Internet "is distributed, it was designed to stop a nuclear attack, or rather, to survive a nuclear attack. If it can do that ... it ought to be able to withstand the U.S. Senate," both assume that the Internet is always redundant, but that

---

56. Whether or not VPNs pilfer public goods is unclear, especially since private corporations now own most of the Internet backbone. VPNs work as follows: first, suppose two geographically separated corporate sites. In order to IP tunnel between them, the firewall for site A would direct all packets (which could be encrypted) to firewall B on the other site (which could authenticate them). Now, through Internet Protocol Security Architecture (IPsec), IP datagrams can be encrypted and an authentication header protocol used to check that nothing or no one has tampered with the packet. In this and other privacy schemes, privacy is conflated with secrecy. And yet, because these signals still function as a language, because as Jacques Derrida argues in "Signature Event Context" (in *Limited Inc.*, trans. Samuel Weber and Jeffrey Mehlman [Evanston, IL: Northwestern University Press, 1988], 1–23), they must be iterable in order to function, they are theoretically decipherable. As well, a secure gateway cannot prevent corruption. Indeed, if one wanted to bring a VPN down, one could simply consistently alter encrypted packets.

specific, popular implementations of TCP/IP are not.[57] They also assume policy-free routing—something true of the pre-privatized Internet. To return to the example of a source-routed packet, if a key gateway—say, the gateway between a local area network and the Internet—refuses source-routed packets, those packets will be discarded (many organizations have only one Internet gateway, which is also a firewall). Many firewalls routinely refuse to allow certain traffic, and certain countries, such as Singapore, seek to usher all traffic through government-sanctioned proxy servers (the default setting in Singapore Web browsers). Since traffic out of Singapore would be halted to a crawl if the government checked every transmission, however, the main gateways (in 2001 at least) simply make sure *a* proxy server is used. Thus, one can access *Playboy* by going through a noncensoring proxy. This work-around does not mean that the Internet cannot be censored but that the effectiveness of censorship depends on local configurations and routing protocols—both of which have been dramatically affected by the privatization of the Internet backbone.

Controlling Code

Those against the increasing privatization of the Internet have stressed the significance of its end-to-end design and free software core. Larry Lessig, in particular, has stressed that "code is law"—decisions that once took place at the level of legislation are now taking place at the level of code. Architecture is therefore politics—the early pioneers, who conflated the Internet with freedom and democracy, Lessig argues, took its code and architecture for granted. Yet code, contrary to Lessig's assertion, is not law.[58] It is better than law; it is what lawyers have always dreamed the law to be: an inhumanly perfect "performative" uttered by no one. Unlike any other law or performative utterance, code almost always does what it says because it needs no human acknowledgment (Lessig himself, while declaring code is law, claims that code has supplanted law: code, not law,

57. Mark Godwin, quoted on "Cybersex: Policing Pornography on the Internet," *ABC Nightline*, June 27, 1995.

58. Lawrence Lessig, *Code: And Other Laws of Cyberspace* (New York: Basic Books, 1999), 6.

increasingly "solves" social problems).[59] Moreover, whereas a law's effectiveness depends on enforcement (self- or otherwise), code's enforcement stems from itself. Code can be sidestepped or broken, but only via technological savvy. Code's colonization of the political makes it a battleground for democracy. According to Lessig, the Internet both weakens governmental sovereignty and strengthens it through governmental collusion with corporations: "The invisible hand of cyberspace is building an architecture that is quite the opposite of what it was at cyberspace's birth. The invisible hand, through commerce, is constructing an architecture that perfects control—an architecture that makes possible highly efficient regulation."[60] The commercialization of the Internet, its transformation into a "secure" marketplace, facilitates control and thus regulation: the interests of commerce and governmental regulation coincide perfectly, making the dispute between commercial organizations and the U.S. legislature over the CDA seem a screen for a more profound collusion.

For Lessig, perfect control signals the demise of democracy: corporations or governmental powers can usurp public decision making through code, thereby rendering cyberspace less free than the "real world." In order to ensure democracy, code must not be owned. Lessig contends, "If the code of cyberspace is owned (in a sense that I describe in this book), it can be controlled; if it is not owned, control is much more difficult. The lack of ownership, the absence of property, the inability to direct how ideas will be used—in a word, the presence of a commons—is key to limiting, or checking, certain forms of governmental control."[61] In this argument, Lessig conflates corporate with governmental regulation, transparency with publicity, and cyberspace with market capitalism, rendering invisible the specific decisions that led to the "ownership" of code. To

---

59.   More important, code can be owned and parsed in a manner unprecedented for any other language product. Although one can produce things with (normal) languages, which can be owned for a period of time, no one "owns" the language per se, and your creations need to be readable in order to run (even to talk about language "products" reveals the extent to which computer and biological codes have transformed language).

60.   Lessig, *Code*, 6.

61.   Ibid., 7.

Lessig, code "not owned" is code protected by GNU's Not Unix (GNU) public license or open source. If a program uses the GNU public license (also known as copyleft), others are free to use this code, but they too must make their source code available. Transparency, not actual ownership of either the code or the system it runs on, thus defines "lack of ownership," and transparency grounds political action. "Only when regulation is transparent is a political response possible," contends Lessig.[62] Transparency also guarantees democracy. Open source is "democracy brought to code," Lessig states, because "an open source code system can't get too far from the will of the users without creating an important incentive among some users to push the project a different way. And this in turn means the platform cannot act strategically against its own."[63] "Open source" becomes a (liberal) check to corporate and governmental power, a means by which, for Lessig, "we build a world where freedom can flourish not by removing from society any self-conscious control; we build a world where freedom can flourish by setting it in a place where a particular kind of self-conscious control survives."[64] With Jeremy Bentham–esque optimism, Lessig assumes that readability ensures democracy (those who can read the code will read it and a "good" consensus will emerge) and that open means public, open means common. Also like Bentham, Lessig makes self-conscious control—the internalization of control—the goal (although unlike Bentham, self-conscious control leads to greater freedom). No matter how transparent a system is, though, an invisible hand (of cyberspace) cannot be seen—and this paradox, stemming from Lessig's conflation of cyberspace and marketplace, reveals his project's limits.

Lessig's second book, *The Future of Ideas*, stresses the importance of TCP/IP rather than software applications (the previous quotations an open source are taken from *The Future*). Internet protocols "embedded principles in the Net," writes Lessig, "constructed an innovation commons at the code layer. Through running on other people's property, this com-

---

62.   Ibid., 181.

63.   Lawrence Lessig, *The Future of Ideas: The Fate of the Commons in a Connected World* (New York: Random House, 2001), 68.

64.   Ibid., 5.

mons invited anyone to innovate and provide content for this space. It was a common market of innovation, protected by an architecture that forbade discrimination."[65] To make this argument—that TCP/IP opened (then) state-owned space, and ensured democratic access to the backbone and its source code—Lessig erases other key issues, such as the influence of academia's "open" structure of knowledge on Internet development, the relatively novel concept of software as a commodity, and restricted access to "end machines" (commercial gateways may make discriminatory routing decisions, but they also enable greater access). Lessig, like John Stuart Mill, also assumes control and innovation are inversely correlated:

The architecture of the original Internet minimized the opportunity for control, and that environment of minimal control encourages innovation. In this sense the argument is linked to an argument about the source of liberty on the original Internet. At its birth, the Internet gave individuals great freedom of speech and privacy. This was because it was hard, under its original design, for behavior on the Net to be monitored or controlled. And the consequence of its being hard was that control was rarely exercised. Freedom was purchased by the high price of control, just as innovation is assured by the high prices of control.[66]

According to Lessig, content and code are parallel systems: the increasing commercialization of networks endangers freedom at both levels by implementing easier control mechanisms and rendering the architecture less democratic (but again, the commercialization of the Internet has led to more democratic access). Remarkably, the assumption that control was rarely exercised because it was hard to do so and that control is antithetical to freedom and innovation overlooks the very operations of TCP/IP (Transmission *Control* Protocol/Internet Protocol).

Alex Galloway, in his analysis of TCP/IP and the bureaucratic structures supporting protocol development, reveals this glaring paradox: "The exact opposite of freedom, that is control, has been the outcome of the last forty years of developments in networked communications. The founding

---

65. Ibid., 85.

66. Ibid., 140.

principle of the net is control, not freedom. Control has existed from the beginning."[67] Significantly, for Galloway, protological control is "a different type of control than we are used to seeing. It is a type of control based in openness, inclusion, universalism, and flexibility. It is control borne from high degrees of technical (organization), no this or that limitation on individual freedom or decision making (fascism)." And so, a "generative contradiction" produces open technology: "In order for protocol to enable radically distributed communications between autonomous entities, it must employ a strategy of universalization, and of homogeneity. It must be anti-diversity. It must promote standardization in order to enable openness," Galloway remarks.[68] Computer protocols do not tolerate deviations—if not followed exactly, compatibility problems will (and often do) occur. If protocols are "antidiversity" because they rely on a common language, however, what entity/system is not antidiversity? What do we mean by diversity? Also, is freedom the exact opposite of control? What precisely is the relationship between medium and content?

Galloway does not simply condemn protological logic, for "it is *through* protocol that we must guide our efforts, not against it."[69] Resistance, like control, is generated from *within* the protological field. He thus turns to tactical media as an effective means of exploiting the "flaws in protological and proprietary command and control, not to destroy technology, but to sculpt protocol and make it better suited to people's real desires. Resistances are no longer marginal, but active in the center of a society that opens up in networks."[70] Galloway's insistence that resistance

---

67. Alex Galloway, "Institutionalization of Computer Protocols," *nettime*, ⟨http://amsterdam.nettime.org/Lists-Archives/nettime-l-0301/msg00052.html⟩ (accessed May 1, 2004). Galloway's critique overlooks Lessig's contention that freedom comes from self-conscious control rather than total lack of it, however this contention does get muted in Lessig's second book.

68. Ibid.

69. Alex Galloway, "Protocol, or, How Control Exists after Decentralization," *Rethinking Marxism* 13, nos. 3/4 (Fall/Winter 2001), 88.

70. Alex Galloway, "Tactical Media and Conflicting Diagrams," *nettime*, ⟨http://amsterdam.nettime.org/Lists-Archives/nettime-1-0301/msg00047.html⟩ (accessed September 13, 2003).

and control constitute, rather than limit, the protological system is crucial, but his notion of sculpting protocol to people's real desires is problematic. As I discuss in more detail in chapters 3 and 5, the relation between technology and desire is highly mediated and slightly paranoid: "people's desires" too are generated by the system. More important, control and freedom are not opposites but different sides of the same coin: just as discipline served as a grid on which liberty was established, control is the matrix that enables freedom as openness. There is, in this sense, no paradox, but there is still a question of freedom—of a rigorous sense of freedom, of freedom, as Jean-Luc Nancy argues, as an experience. In contrast to Lessig and Bentham, publicity, understood as open publication, is not democracy. (Bentham viewed open publication as key to the Panopticon, the disciplinary mechanism par excellence: the only way a Panopticon owner could lose his franchise was by failing to publish his records.) Jodi Dean in *Publicity's Secret: How Technoculture Capitalizes on Democracy* maintains that electronic versions of publicity undermine democracy by magnifying distrust and antagonism rather than rational public discourse. Publicity, she asserts, is the ideology of technoculture; it creates conspiracy theorists and celebrity subjects.

Openness may itself not be democracy, but the openness enabled by communications protocols can point toward this other freedom. Free software, for instance, is not autonomous but creates a structure of sharing. Open source, with its use of an extended creator base made possible by the Internet, pushes this structure further. As well, open source and free software, by belonging to no one, makes democratic struggle possible, makes their code functionally analogous to a public place. As elaborated in more detail in chapter 3, at the heart of democracy lies an empty space: Claude Lefort in *Democracy and Political Theory* argues that because public space belongs by rights to *no one*, because this space cannot be conflated with the majority opinion that may emerge from it, it guarantees democracy.[71] If Lefort's main concern, writing in the 1980s, was totalitarianism and the welfare state, I am now, writing at the beginning of the new millennium, concerned with the increasing role of private corporations in

71.   See Claude Lefort, *Democracy and Political Theory*, trans. David Macey (Minneapolis: Minnesota University Press, 1988), 41.

"public space" and language. Lefort and Thomas Keenan citing Lefort leave their readers with a dangling promise on which they do not deliver —namely, that they will return to the fact that specific individuals or corporations can own public space. This question is even more pressing now, because the problem facing us at present is, What happens when the entity seeming to enforce equality and equal rights is the private corporation instead of the state? What happens when democratic disincorporation stems from consumption rather than voting—when equal rights seem mainly to guarantee access to buying, and when, at the same time, bigoted groups such as the Boy Scouts of America are sanctioned as serving public interest? This is not to say that publicity is not possible within privately "owned" spaces. This means, however, that we need to address the relationship between private/public/political and the transformation of the private/public binary to an open/closed one. Shopping malls and city parks may both be public (or perhaps more properly open) spaces, but they are not equal. Open or free software may be nice, but they leave uninterrogated the question of proprietary hardware and structures of inequality that make it impossible for a good number of workers who create hardware to access software, open or not.

Crucially, both free or open source software are not inherently democratic, representative or otherwise. Although these movements and their products are theoretically open to all, participation depends on education, financial security, leisure time, and so forth, and the final decisions on which revisions get included often lie with one person. Linus Torvalds, who makes the final decisions regarding Linux, is arguably a benevolent dictator, and Richard Stallman, the free software guru, is not known for his democratic tendencies. Still, these movements are not inherently undemocratic either—one can easily imagine them operating under a structure of representative or even Athenian democracy (without the exclusion of women and slaves). The Internet opens up possibilities for reimagining democracy and democratic structures. What is crucial, though, is that the "voluntarism" driving these movements and the division of labor that makes then possible be interrogated.

The Power of Touch

Reducing the Internet to a technical protocol and stressing high-tech Orientalism as a tool for navigation elides the importance of racial and gender differences to hardware production. Hardware has traditionally been produced by women: in the 1980s, the women hired in Silicon Valley to assemble circuit boards and other components were predominantly of Asian heritage. In the 2000s, many women are still Asian, but they live on the other side of the Pacific. The Internet became widely used and computers became personal computers through outsourcing, which combined with advances in technology, has led to dramatic decreases in the prices of personal computers. According to the Catholic Agency for Overseas Development's (CAFOD) 2004 report on the personal computer industry, *Clean Up Your Computer*, until the 1990s, most computer companies produced their products "in-house."[72] Spearheaded by Dell's ruthless cost cutting, however, most computer manufacturing has become outsourced to factories in Mexico and China—factories that often fail to meet minimum wage standards in these countries, and that also subject their workers to strip searches, labor practices designed to prevent collective bargaining, unsafe working conditions (health hazards stem from soldering, noise pollution, and chemical baths used to clean computer components), and excessive overtime. Thus, the actual diversity sustaining cyberspace and the Internet far from reflects the utopian claims of Silicon Valley.

Fiber-optic cable manufacturing itself also tells a significant story. As the Federal Reserve Bank of Richmond notes, before the dot-bomb crisis brought fiber-optic cable production to a crawl, 40 percent of all fiber-optic cable reportedly stemmed from one place, Catawba County in western North Carolina. Firms were lured there by its established labor pool (languishing because of declines in furniture and clothing manufacturing), cheap electricity, and in particular, aggressive efforts by county government. For instance, when Alcatel wanted to expand its fiber-optic factory in the early 1990s, "the county bought the land for the factory and sold it to Alcatel, which paid for the purchase out of future tax payments. The city built a sewage treatment plant to accommodate the

---

72. CAFOD, *Clean Up Your Computer*, ⟨http://www.cafod.org.uk/policy_and _analysis/policy_papers/clean_up_your_computer_report⟩ (accessed May 1, 2004), 7.

factory's waste flow."[73] The benefits of such a move were mixed at best: the companies did add to the local tax base, but many of the benefits went elsewhere. As well, when the companies began to retrench, Catawba County was hit hard; at one point in 2002, it had the highest percentage rate change (in unemployment) over the previous twelve months in the nation. Catawba County is currently looking for another industry to move in. Corning, one of the world's largest producers of fiber-optic cable, has shut down many of its U.S. factories and is now moving toward producing LCDs for flat-screen televisions in Taiwan.

But cyberspace, rather than closing off meaningful contact, can inaugurate it; rather than being the source of inequality, cyberspace can be used as a tool to fight it. CAFOD's report circulates on the Internet, and labor activists use the Internet to organize and raise awareness. Used as a means to get people online, it opens up the possibility of the Internet as a form of (disruptive) communications, and the gap between cyberspace and the Internet perhaps creates dissatisfaction and a desire for something more; the Internet's "underdetermined" nature, as Mark Poster argues, enables greater symbolic participation.[74] Many workers in hardware factories also emphasize that they are not against overseas production, but rather the conditions under which they are forced to work. Their product—these computers that we tap on every day—put us in touch with them, and their lives are profoundly impacted by global telecommunications networks, whether they "use" them or not. The popular myth of computing as a hobby, of computers as originating from garages, the clean lines etched into our Ethernet cards, and the plastic surrounding our new machines blind us to this touch and to the labor necessary for the production of these cards, repeaters, motherboards, and switches.[75]

---

73. Federal Reserve Bank of Richmond, *Seeing the Light*, ⟨http://www.rich .frb.org/pubs/regionfocus/summer03/light.html⟩ (accessed May 1, 2004).

74. See Mark Poster, *What's the Matter with the Internet?* (Minneapolis: University of Minnesota Press, 2001).

75. This touch gives new resonance to Arvind Rajogopal's analysis of the ways in which television in India alters notions of "untouchability" in "Imperceptible Perceptions in Our Technological Modernity," in *New Media Old Media: A History*

Also, the very fictional concept of cyberspace can lead elsewhere. Chela Sandoval, in *Methodology of the Oppressed*, sees cyberspace as emancipating, but not because it enables bodiless exultation. To enter a world where anything is possible is to enter cyberspace—a place already created by the methodology of the oppressed:

It has been assumed that the oppressed will behave without recourse to any *particular* method, or rather, that their behavior consists of whatever acts one must commit in order to survive, whether physically or psychically. This is exactly why the methodology of the oppressed can now be recognized as the mode of being best suited to life under neocolonizing postmodern and highly technologized conditions in the first world; for to enter a world where any activity is possible in order to ensure survival is to enter a cyberspace of being. In the past this space was accessible only to those forced into its terrain. As in [Donna] Haraway's definition above, this cyberspace can be a place of boundless and merciless destruction—for it is a zone where meanings are only cursorily attached and thus capable of reattaching to others depending on the situation to be confronted. Yet this very activity also provides cyberspace its decolonizing powers, making it a zone of limitless possibility.[76]

Sandoval here intertwines necessity and choice: one chooses what is necessary for survival, but by realizing that one can survive, one realizes that any activity is possible. The merciless movement between Signifiers, Signifieds, and referents becomes decolonizing, becomes a necessary violence. This possibility of endless opportunity through the reworking of language and poetic expression is certainly inspiring, but to what extent does the movement between Signifiers, Signifieds, and referents enable a zone of limitless possibility, rather than indicate the growing mutation of language by code? Is anything possible, and what is at stake in marking off cyberspace in this manner? This book responds to these questions in a far

---

and Theory Reader, eds. Wendy Hui Kyong Chun and Thomas W. Keenan (New York: Routledge, 2005), 275–284.

Chela Sandoval, *Methodology of the Oppressed* (Minneapolis: University of Minnesota Press, 2000), 177.

less optimistic manner, examining the impact of "scenes of empower-ment" and "high-tech Orientalism." To be clear, by analyzing the ques-tion of race and cyberspace in this way, I am not dismissing Sandoval's claims; rather, I am framing my analysis within hers to highlight the fact that cyberspace as utopia has a double valence. Not all utopian views of cyberspace are perpetuated with an eye to creating a marketplace of ideas, or through the manipulation of people of color. Yet rather than explore the utopian possibilities of a space in which anything is possible, I argue that by refusing this myth, the Internet can enable something like democracy. Thus this chapter, moving from notions of cyberspace to the workings of the Internet and the bridging of control and freedom has asserted that the conception of the user as an empowered agent must be interrogated, not because users are completely powerless or because their acts are delusional, but rather because vulnerability and a certain loss of control drives communication, drives our using. As I contend in the next chapter, it is by exploring the possibilities opened up by our vulnerabil-ities, rather than by relegating this vulnerability as accidental or porno-graphic, that the question of democracy can be rigorously engaged.

SCREENING PORNOGRAPHY

Cyberporn became a pressing public danger in 1995. The CDA passed the U.S. Senate with an overwhelming majority after senators perused tightly bound printouts of "perverse" images that Senator James Exon's "friend" had downloaded. Subsequently passed by Congress, the Telecommunications Act of 1996 both deregulated the telecommunications industry—allegedly opening access for all citizens to the Internet—and regulated Internet content for the first time. *Time* and *Newsweek* published special features on cyberporn with the respective titles, "On a Screen Near You, Cyberporn: A New Study Shows How Pervasive and Wild It Really Is" and "No Place for Kids? A Parent's Guide to Sex on the Internet." Philip Elmer-Dewitt's "On a Screen Near You, Cyberporn" launched a particularly heated online and off-line debate over pornography's pervasiveness on the so-called information superhighway, and was accused of launching the "Great Internet Sex Panic of 1995."[1] Taking its facts from a Carnegie Mellon undergraduate thesis, the *Time* article stressed the fact that 83.5 percent of all Usenet images were pornographic (as Professors Donna Hoffmann and Thomas Novak quickly countered, however, such a statistic was insignificant—less than .5 percent of all Usenet messages contained pornographic images).[2] Although the methodology and

---

1. See Mike Godwin, "Journoporn: Dissection of the *Time* Scandal," *Hotwired*, ⟨http://hotwired.wired.com/special/pornscare/godwin.html⟩ (accessed May 1, 2004).

2. Donna Hoffmann and Thomas Novak, "A Detailed Analysis of the Conceptual, Logical, and Methodological Flaws in the Article: 'Marketing Pornography

conclusions of *Time*'s featured "Carnegie Mellon Report" were eventually discredited, the special issue served as evidence in the U.S. House debate over Senator Exon's bill, and the furor over cyberporn—based on second-hand sightings and rumors, and central to negotiating private and public power—was not easily dispelled.[3]

In fact, cyberporn hype increased as rumors of a pornographic gold rush followed the Supreme Court's decision to strike down the CDA. If stories of cyberporn's prevalence and danger abounded in 1995, news of its profitability triumphed in 1996–1997, sparking the dot-com craze. On August 20, 1997, the *Wall Street Journal*'s Thomas Weber pronounced, in "The X Files: For Those Who Scoff at Internet Commerce, Here's a Hot Market: Raking in the Millions, Sex Sites Use Old-Fashioned Porn and Cutting-Edge Tech—Lessons for the Mainstream," "find a web site that is in the black, and, chances are, its business and content are distinctly blue."[4] Weber stressed that nonpornographic commercial Web sites could learn much from X-rated sites' strategies, such as placing notices in appropriate discussion groups, running ads on search engines, and most importantly, zapping encrypted credit card numbers despite widespread fears

on the Information Superhighway,'" ⟨http://elab.vanderbilt.edu/research/topics/cyberporn/rimm.review.htm⟩ (accessed May 1, 2004).

3.   The perceived importance of this article is truly remarkable. As well as being used as evidence and routinely referenced as *the* spark that launched the porn panic, it also provoked a series of articles in magazines, such as *Harper's* and *Hotwired*, and newspapers, such as the *Boston Globe*, in which the errors of *Time* were documented, and a steady stream of articles in which Internet novices took to the Internet superhighway in the footsteps of the little boy figured on *Time*'s cover (see Andy Smith, "Okay ... Where's the Cyberporn?" *Providence Journal*, July 16, 1995, 1E.). The report on which the *Time* article was based, "Marketing Pornography on the Information Superhighway," was the senior thesis of Marty Rimm, an engineering student at Carnegie Mellon University. After its publication, several advisers associated with the report withdrew their names. For more on the controversy surrounding this report, see ⟨http://129.59.210.73/cyberporn.debate.cgi⟩.

4.   Thomas Weber, "The X Files: For Those Who Scoff at Internet Commerce, Here's a Hot Market: Raking in the Millions, Sex Sites Use Old-Fashioned Porn and Cutting-Edge Tech—Lessons for the Mainstream," *Wall Street Journal*, August 20, 1997, A1.

of hackers (the CDA essentially endorsed porn sites willingness to zap sites' credit cards as socially responsible). Apparently, someone was listening, and that same year, a *CNN/Time Impact* special on cyberporn documented secret consultations between cyberporn Web mistresses/masters and corporate Web site designers. The successes of cyberporn sites, the special (in the form of CNN's Bernard Shaw) argued, were convincing American corporations that users would use their credit cards online and users that transmitting these numbers did not guarantee catastrophe. Thus, corporations like IBM could only offer strategies to "work the Web" (as its 1997 television commercials declared) after a private detour through cyberporn/sex. Perhaps it is no accident that the name of one of the most popular Web sites and *the* Web site to which academics first submit their credit card numbers, amazon.com (which had its first profitable year in 2004), could easily be that of a porn site.

Pornography therefore was, and still is, central to the two issues that map the uneasy boundary between public and private: regulation and commerce. The Internet's privatization paved the way for cyberporn to the extent that it made digital pornography a hypervisible threat/phenomenon, and cyberporn paved the way for the "Information Superhighway" to the extent that it initiated the Internet gold rush and caused media, government, and commercial companies to debate seriously and publicly the status of the Internet as a mass medium. Prior to the Internet going public (by being taken over by private corporations), legislators showed no concern for minors who accessed the alt.sex hierarchy or "adult" BBSs; pornography's online presence was so well-known among users it was not even an open secret. "Discovering" the obvious, the media and politicians launched a debate about "free" speech focused on assessing, defining, and cataloging pornography.

Pundits and critics responded to the initial "shock" produced by these cyberporn headlines with clichés and/or dubious historical truisms. Disparate sources from CNN's Shaw to online pornographers explained cyberporn's profitability with the adage "sex sells." Catharine MacKinnon, interpreting the "Carnegie Mellon Report" as vindicating her own pro-censorship position, wrote, "Like a trojan horse, each new communications technology—the printing press, the camera, the moving picture, the tape recorder, the telephone, the television, the video recorder, the VCR, cable, and, now, the computer—has brought pornography with it.

Pornography has proliferated with each new tool, democratizing what had been a more elite possession and obsession, spreading the sexual abuse required for its making and promoted through its use."[5] The slogans "sex sells" and "technology is a pornographic Trojan horse" assume that pornography, sexual practices, and commerce have not changed since the printing press, or since early prostitution, and that the medium makes no difference. Questions of historical specificity aside, these truisms ignore the recurring representation of mass media *as* pornographic inundation, even as they link pornography to mass markets and mass media (historians Lynn Hunt, Lynda Nead, and Robert Darnton, among others, have investigated the relationship between the three more rigorously).[6] Lastly, these slogans miss the interlinked importance of cyberporn commerce and regulation. As mentioned previously, the government's attempt to regulate pornography led to the pornographic gold rush.

Michel Foucault's argument in *The History of Sexuality* that sexuality is instrumental to power relations could also be wielded to dispel surprise at cyberporn's role in market capitalism and government regulation. Cyberporn, as yet another "new technology of sex," predictably "require[s] the social body as a whole, and virtually all of its individuals, to place themselves under surveillance."[7] Although Foucault's analysis helps elucidate cyberporn, treating cyberporn as simply more evidence to support Foucault's initial theses on the *history* of sexuality forecloses the present and blunts the future. It also erases key transformations in surveillance. For whom and to whom are we to place ourselves under surveillance? What

---

5.   Catharine MacKinnon, "Vindication and Resistance: A Response to the Carnegie Mellon Study of Pornography in Cyberspace," *Georgetown Law Review* 93, no. 5 (June 1995): 1959–1967.

6.   See Lynn Hunt, ed., *The Invention of Pornography: Obscenity and the Origins of Modernity, 1500–1800* (New York: Zone Books, 1993); Lynda Nead, *The Female Nude: Art, Obscenity, and Sexuality* (New York: Routledge, 1992); and Robert Darnton, *The Forbidden Best-Sellers of Pre-Revolutionary France* (New York: W. W. Norton, 1995), and *The Literary Underground of the Old Regime* (Cambridge: Harvard University Press, 1982).

7.   Michel Foucault, *The History of Sexuality, Volume 1: An Introduction*, trans. Robert Hurley (New York: Vintage Books, 1978), 116.

exactly does surveillance comprise, and how does it impact our actions? According to Foucault, sex moved from a religious to a secular concern as power moved from monarchal to disciplinary power during the eighteenth century.[8] Sex's virtual reemergence as a "new" public concern exposes failures in traditional forms of disciplinary power and the emergence of control-freedom.

In this chapter, I examine cyberporn's role in addressing and negotiating the "public" as guardians of underage users (and thus guardians of "our future") as well as a community of possible users. Although pro- and anti-CDA forces mainly sparred over competing notions of the Internet as a marketplace, discussions about cyberporn also facilitated understandings that went beyond these notions, for in these discussions, electronic interchanges were acknowledged as contagious and exposed. That is, they were discussed in terms that engaged the dangers and freedoms of democracy, albeit in terms that made vulnerability contingent rather than constituent. Those arguing for Internet regulation as a necessary "civilizing" step saw no difference between the Internet's pornographic content and the Internet itself. To them, the medium was the message, and the message was the pornographic invasion of the home. As drastic as this sounds, this understanding actually ignored the Internet's ramifications, for it assumed that the Internet's effect depended on *content*. Yet as this chapter argues, if the Internet enables communication and transforms the home, it does so independent of content. So if Alexis de Tocqueville once commended Americans for taming the dangers of freedom through their enjoyment, the Internet is forcing Americans to revisit, if not renegotiate, the joy of freedom and the relationship between governmental and self- (that is, corporate-) censorship.

Exposed, or the Walls of the Home Cannot Hold

As noted above, 1995 marked the emergence of cyberporn into the public eye mainly through an extraordinarily controversial and influential *Time* article (based on an equally controversial report/undergraduate thesis by then Carnegie Mellon senior Marty Rimm). In the article, author

---

8.   Ibid.

Elmer-Dewitt argued that public furor over cyberporn exposed a peculiar paradox: "sex is everywhere," and yet "something about the combination of sex and computers seems to make otherwise worldly-wise adults a little crazy."[9] Specifically, "most Americans have become so inured to the open display of eroticism—and the arguments for why it enjoys special status under the First Amendment—that they hardly notice it's there," yet online pornography, which most Americans had never viewed in July 1995, had become hypervisible.[10] Indeed, secondhand viewings of cyberporn were sufficient to pass legislation and raise public concern. According to a 1995 survey run by Princeton Survey Research Associates, 85 percent of people polled were concerned about children seeing pornography on the Internet.[11]

The types of images that Senator Exon provided determined his "blue book's" efficacy. As Elmer-Dewitt relates, "Exon had asked a friend to download some of the rawer images available online. 'I knew it was bad,' he [Exon] says. 'But then when I got on there, it made *Playboy* and *Hustler* look like Sunday-school stuff.'"[12] Materials available on the Internet, claimed Elmer-Dewitt, "can't be found in the average magazine rack: pedophilia (nude pictures of children), hebephilia (youths) and what the researchers call paraphilia—a grab bag of 'deviant' material that includes images of bondage, sadomasochism, urination, defecation, and sex acts with a barnyard full of animals."[13] Indeed, Rimm, in order to classify the pornography that his research team supposedly found, added the following to the standard Dietz-Sears categorizations of pornography: incest, "amazing," pedo/hebephile, dog style, swing, whore, hair color, obese,

---

9.   Philip Elmer-Dewitt, "On a Screen Near You, Cyberporn: A New Study Shows How Pervasive and Wild It Really Is," *Time*, July 3, 1995, 38.

10.   Ibid., 38.

11.   Quoted in "Cybersex: Policing Pornography on the Internet," *ABC Nightline*, June 27, 1995.

12.   Elmer-Dewitt, "Screen," 42.

13.   Ibid., 40. Many commentators argued that although these images could not be found on the average magazine rack, they did not compare with those available in specialty shops; many of the images online were low-quality scans.

muscular, shower, outdoor, petting, panties, Asian, interracial, portraits, famous models, and emotions.[14] These categories resolve sexuality more finely, loosening it from object-choice gender and thus upsetting mainstream notions of "sexual orientation." According to Eve Kosofsky Sedgwick, the gender of one's object choice "emerged from the turn of the century, and has remained . . . *the* dimension denoted by the now ubiquitous category of 'sexual orientation.'"[15] This upsetting, combined with images of necrophilia and pedophilia and extreme descriptions (oral sex described as "choking"), shut down any conversation about pornography's value (although they did not agree on how to deal with such materials, those for and against the CDA agreed that they were bad). Also, as exemplified by Exon's blue book's success, the "deviance" and accessibility of this obscene material belied the need to prove its pervasiveness. Yet unlike print pornography, which usually resides in specially marked places, online pornography can be accessed by any working networked computer equipped with the proper software. Thus, online pornography can be pervasive without being extensive: theoretically, there could be only one "deviant" pornography site, but this one site could make its material available simultaneously to multiple users.

Elmer-Dewitt, ignoring this new relationship between pervasiveness and extensiveness, maintains that the prevalence of deviant pornography stems from context and reveals fundamental truths about "ourselves":

Pornography is different on the computer networks. You can obtain it in the privacy of your own home—without having to walk into a seedy bookstore or movie house. You can download only those things that turn you on, rather

---

14.   Marty Rimm, "Marketing Pornography on the Information Superhighway: A Survey of 917,410 Images, Descriptions, Short Stories, and Animations Downloaded 8.5 Million Times by Consumers in over Forty Countries, Provinces, and Territories," *Georgetwon Law Journal* 83, no. 5 (June 1995): appendix A. Importantly, Rimm's team did not actually download and examine all the images it accounted for, but depended to a large extent on the descriptions of these images. Given this, many of these images were probably redundant.

15.   Eve Kosofsky Sedgwick, *Epistemology of the Closet* (Berkeley: University of California Press, 1990), 8.

than buy an entire magazine or video. You can explore different aspects of your sexuality without exposing yourself to communicable diseases or public ridicule. (Unless, of course, someone gets hold of the computer files tracking your online activities, as happened earlier this year to a couple dozen crimson-faced Harvard students.)[16]

Hence, the "truth" revealed by Rimm's "Marketing Pornography" thesis is that in private, without fear of contamination or exposure, sexuality veers toward the deviant; technology brings to the surface the perversity lying within us all. To assert this, both Elmer-Dewitt and Rimm assume that the pornography *you* download corresponds to *your* sexuality. As I argue more fully later, however, there is an important gap between download and identity, between looking and acting. The thrill of downloading so-called deviant pornography stems from both the content and the very act of searching and downloading "blasphemous knowledge."

The importance of context rather than content means the supposed perversity peculiar to electronic pornography spreads to all online material, for all materials—pornographic or not—can be read with the same belief that alone before our personal computers, we temporarily evade public norms. Indeed, this whole discussion, framed as a personal interaction between an individual and one's monitor, reveals the computer industry's success in reconceiving computers as personal belongings rather than institutional assets. The computer as personal or private is deceptive, though, since the possibility of someone getting "hold of the computer files tracking your online activities" is constitutive of, rather than accidental to, this medium. That surfing the Web has been heralded as an act cloaked in secrecy that needs to be publicly illuminated is simply bizarre, and file tracking is acknowledged by the many users whose behavior does change in front of a personal computer, albeit not in the terms Elmer-Dewitt describes. Those aware of and concerned with tracking treat possibility as fact, and assume that all their electronic data transfers are recorded and analyzed—an assumption that flies in the face of their everyday experience with crashing computers, undelivered e-mail messages, and inaccessible Web sites. They therefore encrypt their messages, guarantee-

---

16.   Elmer-Dewitt, "Screen," 40.

ing that their messages will be recorded.[17] Those who know of, but are not concerned with, tracking believe they can "survive the light" because they either consider the likelihood of exposure negligible, or think the standards for public interactions online different, or want their misdemeanors to be spectacular. Regardless, visibility fails to produce automatically disciplined subjects (if it ever did).[18]

Visibility's failure to ensure discipline underscores the differences between the Internet and the Panopticon. The Internet may enable surveillance, or "dataveillance," but it is not a Panopticon. First, computer networks "time shift" the panoptic gaze; second, users are not adequately isolated. According to Bentham, the inmate had to "conceive himself to be [inspected at all times]" in order for the Panopticon to work.[19] The inspector's quick reaction to misbehavior early in the inmate's incarceration and the central tower's design, which made it impossible for the inmate to verify the inspector's presence, were to make the inmate internalize the gaze and reform. Contrary to Hollywood blockbusters, real-time spying is the exception rather than the norm on computer networks—and arguably all media frustrate real-time discipline. Digital trails and local memory caching, inevitably produced by online interactions, do make prosecution easier, though.[20]

---

17.   According to Whitfield Diffie and Susan Landau in their description of NSA intercept machines, "If an intercepted message is found to be encrypted, it is automatically recorded. This is possible because at present only a small fraction of the world's communications are encrypted" (*Privacy on the Line: The Politics of Wiretapping and Encryption* [Cambridge: MIT Press, 1998], 91).

18.   Michel Foucault himself calls Jeremy Bentham's belief that opinion was always "good" an optimistic illusion: the utilitarians "overlooked the real conditions of possibility of opinion, the 'media' of opinion, a materiality caught up in the mechanisms of the economy and power in its forms of the press, publishing and later the cinema and television" ("The Eye of Power," in *Power/Knowledge: Selected Interviews and Other Writings, 1972–1977*, trans. Colin Gordon [New York: Pantheon Books, 1980], 161–162).

19.   Jeremy Bentham, "Panopticon; or, the Inspection-House," in *The Panopticon Writings*, ed. Miran Božovič (London: Verso, 1995), 34.

20.   As stated in the introduction, in the United Kingdom and the United States, law enforcement officers do not need a search warrant to determine the sending and receiving locations of one's e-mail but do need one to read e-mail.

Thus, it is not that someone could be looking but that—at any point in the future—someone *could* look. In front of one's "personal computer," one does not immediately feel the repercussions of one's online activities, but one is never structurally outside the gaze (yet to be precise, there is no gaze, since the function of seeing has been usurped by reading and writing—seeing has become increasingly metaphoric in the age of fiber optics). Whether or not someone will or can access your files, however, is fundamentally uncertain and depends on software. As I discussed in the introduction, glossing over this uncertainty and this dependence screens information's ephemerality and software's impact.

Significantly, computer network-aided interactions validate so-called deviant behavior, tempering the effect of time-shifted visibility. Although Bentham, to offset solitary confinement's detrimental effects, revised his plan so that two or three quiet inmates could work together, his system depended on isolating the inmate. In contrast, the Internet physically separates, but virtually connects. The Department of Justice in its portable guide to law enforcement officers, *The Use of Computers in the Sexual Exploitation of Children*, argues, "communicating with other people who have similar interests validates the offender's interests and behavior. This is actually the most important and compelling reason that preferential sex offenders are drawn to the online computer."[21] On the Internet, others mirror one's perversities; one recognizes in others one's personal idiosyncrasies. On the Internet, one becomes a statistic, but through this reduction, one's "personality" is reinforced (statistical analysis itself is predicated on individuality—one needs prediction only when one is uncertain of a result).

This validation of so-called private desires, also described as community, leads to broader questions of computer networks and crises of discipline—questions foreclosed by the emphasis on children. This myth of the agentless child victimized by cyberporn "simplifies" issues (just as focusing on pornographic materials simplifies the issue of electronic exposure); it enables adults to address issues of vulnerability without acknowl-

21.   U.S. Department of Justice, *The Use of Computers in the Sexual Exploitation of Children*, Office of Justice Programs, 1999, 4.

edging their own, and enables parents to admit their deficiencies as disciplinary agents without fear of condemnation. As congressman Pete Geren
put it, "For many of us our children's knowledge of the computer, just—
to say it dwarfs ours is not really an exaggeration at all."[22] In hearings and
articles about cyberporn, lawmakers and others transform their own anxieties into concern over their children's (and thus our future generation's)
sexual safety, whether or not they actually have said (computer-savvy) children. Online pornography intrudes into the home, circumventing the normal family disciplinary structure, subjecting children and threatening to
create deviant subjects. As Elmer-Dewitt asserts, "This [exposure to
cyberporn] is the flip side of Vice President Al Gore's vision of an information superhighway linking every school and library in the land. When
kids are plugged in, will they be exposed to the seamiest sides of human
sexuality? Will they fall prey to child molesters hanging out in electronic
chat rooms?"[23] Similarly moving without explanation from online pornography to child molestation, CDA proponents proffered "high-profile
cases" of child abduction/seduction to support their demand that Internet
content be regulated in the same manner as television.[24]

Elmer-Dewitt himself favors nonlegislative means to contain this
"out-of-control" medium. Although parents have "legitimate concerns
about what their kids are being exposed to," for Elmer-Dewitt, turning
off the light is not the answer. Rather, he suggests, "Men and women
have to come to terms with what draws them to such images. Computer
programmers have to come up with more enlightened ways to give users
control over a network that is, by design, largely out of control."[25]

In the end, suggests Elmer-Dewitt, this breach must be stopped by
renewed family discipline rather than state regulation:

---

22.   Quoted in House of Representatives, *Cyberporn: Protecting Our Children from
the Back Alleys of the Internet*, 104th Cong., 1st session, (Washington, DC: U.S.
Government Printing Office, 1995), 80.

23.   Elmer-Dewitt, "Screen," 40.

24.   Ibid., 40–42.

25.   Ibid., 40.

Pornography is powerful stuff, and as long as there is demand for it, there will always be a supply. Better software tools may help check the worst abuses, but there will never be a switch that will cut it off entirely—not without destroying the unbridled expression that is the source of the Internet's (and democracy's) greatest strength. The hard truth, says John Perry Barlow, co-founder of the EFF [Electronic Frontier Foundation] and father of three young daughters, is that the burden ultimately falls where it always has: on the parents. "If you don't want your children fixating on filth," he says, "better step up to the tough task of raising them to find it as distasteful as you do yourself."[26]

The hard truth behind cyberporn, then, is simple: given that pornography will thrive as long as there is a demand for it, parents must ensure that their children are not attracted to filth.[27] The emphasis moves from the medium to the message.

Although Elmer-Dewitt's article ends by endorsing family discipline, its accompanying illustrations undercut this resolution by emphasizing the computer connection as breach. These illustrations do not reproduce on-line pornography but rather play with the tension between exposure and enlightenment, between licit and illicit knowledge. They reveal that the

---

26.   Ibid., 45.

27.   Elmer-Dewitt's resolution of this concern, the delegation of censorship and discipline to the family, is a familiar argument. Advocates against television censorship have consistently maintained that if parents are concerned about violence on television, they should watch television with their children, rather than using it as a cheap substitute for babysitting. This position assumes a clear demarcation between public and private, display and consumption, government and family. It also assumes that consumption drives the production of pornography, and that the family is responsible for regulating and producing sexual desire. The popularity of "deviant" pornography, then, points to familial failures or intimate truths about "ourselves" (as consumers of pornography). In order for the Internet to be a public space in which one may make public use of one's reason, so this position goes, it must be treated as a free marketplace of ideas. Similarly, in order for the Internet to remain democratic, the family, rather than the government, must enlist new and more intrusive disciplinary techniques, must take on the task of supervising their children's online activities, rather than protesting ignorance.

| Figure 2.1 |

Cover of *Time's* July 3, 1995, issue

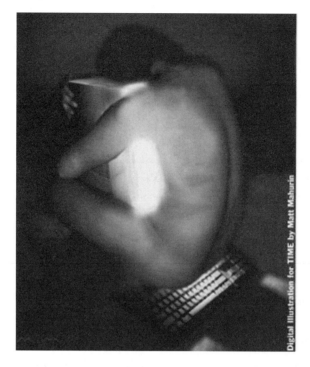

Digital Illustration for TIME by Matt Mahurin

| Figure 2.2 |
Time illustration

fear of too much light, of too much exposure and uncontrollable contact, is what makes worldly wise adults a little crazy, if not paranoid. The cover of *Time*'s cyberporn special issue (figure 2.1) enacts first contact. The glare of the computer screen, in stark contrast to the darkened room, simultaneously lights up and casts shadows over the startled blond boy's face, literalizing his enlightenment/overexposure. His eyes and mouth are open, and his tiny hands are lifted off the keyboard in horror or surprise: the images emanating from his monitor open and immobilize his facial orifices. The roundness of his open mouth evokes images of vagina-mouthed inflatable dolls. Further, the screen's glare exposes wrinkles under the little boy's eyes, signs of premature aging, of a loss of innocence that belie his tiny hands and two front teeth. His solitude in front of the computer screen and the room's dim lighting suggest secrecy. Instead of basking in the cozy light of his family home, he is immobilized by *us* watching him, since we—the readers—are in the position of the intruding pornographic image. Or else he serves as our mirror image, his surprise and invasion mirroring our own. This image evidences—through a mass-circulated print representation—the spectacularity of Internet pornography and, by implication, the Internet as a whole.

The full-page illustration (figure 2.2) that introduces this article features the screen's glare more prominently. An anonymous, presumably male figure wraps his arms and legs tightly around the bright computer monitor, his bottom resting gingerly atop a lighted keyboard. Again, the computer screen serves as the room's only source of light, and this bright light shines through his translucent body. If the cover emphasized the innocence of the little boy, this image represents the possible ramifications of first contact: the desire to be touched and penetrate/be penetrated. In a logic akin to Catharine MacKinnon's in *Only Words*, the progression of these images implies that pornography creates a pornographic culture by inciting desires/erections in its viewers rather than revealing the perversity lurking within us.[28] As opposed to MacKinnon, however, these images seem to support pro-CDA arguments that pornography's danger stems not from its abuse of women but rather from its violation of viewers, specifically its underage viewers, who are unable to reason against its temptations.[29] Also, rather than desiring the images on the screen, or more

---

28.   According to Catharine MacKinnon, pornography is "constructing and performative rather than … merely referential or connotative. The message of these materials, and there is one, as there is to all conscious activity, is 'get her,' pointing at all women, to the perpetrators' benefit of ten billion dollars a year and counting. This message is addressed directly to the penis, delivered through an erection, and taken out on women in the real world" (*Only Words* [Cambridge: Harvard University Press, 1993], 21). In terms of its effect on women, MacKinnon notes that "as Andrea Dworkin has said, 'pornography is the law for women.' Like law, pornography does what it says. That pornography is reality is what silenced women have not been permitted to *say* for hundreds of years" (41).

29.   If, as Lynne Segal argues, "feminists could indeed rightly claim it as a victory that whereas once the concern about pornography was mainly over its effects upon those who consumed it, today the concern is mainly over its effects upon those who are represented by it" ("Does Pornography Cause Violence? The

properly the objects represented by these images, this anonymous man appears to desire the computer itself, highlighting another "perversity" or "obscenity" associated with online pornography. This image mirrors Jean Baudrillard's complaint: "The obscenity of our culture resides in the confusion of desire and its equivalent materialized in the image; not only for sexual desire, but in desire for knowledge and its equivalent materialized in 'information.'"[30] Paul Virilio, in *Open Sky*, similarly argues that cybersex is a form of prophylaxis that threatens the health of the human species. Cyberporn seems to amplify pornographic images' tendencies to "usurp their referent."[31] Desire detours through the transportation medium, posing the following questions: Does the viewer of cyberporn desire the computer, the image, or the image's referent, if such a referent exists? Can these objects of desire be separated? These illustrations allege that violation leads to contagious and perverse desire, that pornography starts a wildfire that overwhelms and engulfs enlightenment and reality, so that content cannot be separated from medium.

These images portray pornography as unsupervised enlightenment, as information that perverts rather than advances; the 105th U.S. Congress, quoting Dr. Gary Brooks, also advanced this claim in its report on the Child Online Protection Act (COPA) or CDA II:

Search for Evidence," in *Dirty Looks: Women, Pornography, Power*, eds. Pamela Church Gibson and Roma Gibson [London: British Film Institute, 1993], 11), the move to condemn pornography because of its potential effects on its viewers rather than on those represented is a step backward for procensorship feminists. This step backward stems from the mechanics of online pornography: since an electronic image can be easily manipulated, since there is not necessarily a referent, procensorship forces could no longer claim that pornography always abused the women it represented. In response to this loss of indexicality, lawmakers amended child pornography laws in 1996 to include images that simulated persons under eighteen.

30.   Jean Baudrillard, *The Ecstasy of Communication*, trans. Bernard and Caroline Schutze (Brooklyn, NY: Semiotext(e), 1988), 35.

31.   For more on the complicated relationship between pornographic images and their referents, see Lucienne Frappier-Mazur, "The Truth and the Obscene Word in Eighteenth-Century French Pornography," in *The Invention of Pornography*, ed. Lynn Hunt, 203–221.

The type of information provided by pornography … does not provide children with a normal sexual perspective. Unlike learning provided in an educational or home setting, exposure to pornography is counterproductive to the goal of healthy and appropriate sexual development in children. It teaches without supervision or guidance, inundating children's minds with graphic messages about their bodies, their sexuality, and those of adults and children around them.[32]

Pornography, and by extension the Internet, inundates and overwhelms. According to those arguing for the CDA, *the message is the medium*: it (the Internet as pornography) threatens the future health of society by enabling unsupervised enlightenment, a situation that Brooks associates with learning outside normal disciplinary settings. And so the solution is to change the message and thus the medium.

Further complicating this "obscene" and risky scenario, the illustration on *Time*'s table of contents page reverses the gaze (figure 2.3). Although the caption reads "People are looking at pictures of *what* on the Internet?" an eye peers from the monitor at the viewer. Analogous to Elmer-Dewitt's use of "you," this image places *Time*'s reader in front of the monitor, suggesting that everyone is at risk. Once more, the screen provides the only light source and the eye appears wrinkled, indicating an aged other, prematurely so or not. The computer screen becomes a window through which this other looks at and exposes us. Rather than an interface, the screen becomes an intraface: a moment of face-to-face contact with this mature eye. The monitor monitors: someone could be watching.

Building on this ambiguity between watcher/watched, *Time*'s other illustrations make explicit Internet pornography's "deviance." In the one illustration that is clearly an "artist's conception," a man, hiding behind a computer screen, lures a little child with a bright red lollipop (figure 2.4). This image alludes to oral sex or homosexual contact, but more important, it illustrates one of society's hypervisible fears: young boys being lured by older men (presumably, this scenario is so dangerous that it could only be

---

32. U.S. Congress, "105th Congress Report" (105th Cong., 2nd Session) 105–775 Child Online Protection Act ⟨http://www.epic.org/free-speech/censorship/hr3783-report.html⟩ (accessed August 1, 2004).

Cover: People are looking at pictures of *what* on the Internet?

| Figure 2.3 |
Time illustration

rendered as a drawing rather than a digitally altered photograph). This picture of a vulnerable prepubescent child clashes with empirical evidence: most online seduction/abduction cases involve adolescents rather than young children hanging out in AOL tree houses. As the Department of Justice explains, "Investigators must recognize that children who have been lured from their homes after online computer conversations were not simply duped while doing homework. Most are curious, rebellious, or troubled adolescents seeking sexual information or contact."[33] The group

---

33.   U.S. Department of Justice, *The Use of Computers*, 6.

Digital Illustration for TIME by Matt Mahurin

| Figure 2.4 |
Time illustration

"at risk" for statutory rape are "adolescent boys who spend many hours 'hacking' on their computers," adolescents who have wills and desires that others are constantly trying to deny them (unless, of course, they commit a serious crime, at which point the same cultural conservatives who argue that "children" are victimized by cyberporn, contend that minors should be tried as adults).[34] The myth of the agentless child is precisely that, and through this myth, risks endemic to all online interactions are refigured as catastrophic risks to children, and homosexuality is rewritten as a form of child abduction.

As mentioned earlier, law enforcement and other psychological "experts" link the Internet to pedophilia because it enables "community," not because it enables greater access to children (indeed, many articles insist that real-life access to children is much easier—pedophiles tend to be schoolteachers, coaches, and priests).[35] It also facilitates the circulation of child pornography, which some view as a precursor to, and others a substitute for, the actual pedophiliac act.[36] Regardless, proactive police units use the hypervisible pedophile to validate their methods, which arguably—although ostensibly not illegally—entrap: police officers pose as curious young boys and actively seek, and perhaps create, pedophiles.[37]

34. Ibid., 5.

35. As one convicted pedophile put it, "On the computer, the search for a victim is an arduous task that's fraught with danger due to the intensity of law enforcement.... Besides ... victims are too easy to find in other places.... [Successful pedophiles] are better with your children than you are. They give them more attention. They are your swim coach, your Sunday school teacher—people you trust to come into contact with your child every single day" (quoted in Bob Trebilcock, "Child Molesters on the Internet," *Redbook*, April 1, 1997, 102).

36. Paul Virilio in *Open Sky* (trans. Julie Rose [London: Verso, 1997]) views cybersex more generally as a substitute for sex and thus a form of species suicide.

37. For a provocative case regarding the relationship between Internet regulation, surveillance, and pornography, see Laura Kipnis's analysis of *United States v. Depew*. In this case, two men (one a pedophile and the other a sadomasochistic "top") were contacted over the Internet by an undercover San Jose, California, police officer who suggested they make a snuff film. Although no child was ever kidnapped or killed, Depew himself withdrew from the project, and the lines between fantasy and intent were extremely difficult to draw, Depew was sentenced

The so-called "Innocent Images" initiative, which simultaneously seeks to make images innocent and declares no image innocent, foreshadows the proactive police methods implemented more broadly after September 11, 2001. These techniques that lead to the routine arrest and conviction people for crimes they did not actually but "intended" to commit (intension becomes the point of contention), reveal the difference between discipline and control. To be clear, I am not arguing for child pornography or pedophilia but rather seeking to understand how, through the Internet, pedophilia has been established as *the* most hypervisible deviant sexuality useful to methods of control. Significantly, the popular conception of the Internet as aiding and abetting pedophiles does not reflect networking protocols but rather propaganda about the Internet as "empowering" and anonymous. Given its constitutive tracking ability, the Internet could easily have been heralded as facilitating the *prosecution* of pedophiles (that is, now we can catch those people who, before the Internet, circulated their images in a less accessible manner). The fact that it was not so heralded reveals assumptions about technology as *inducing* "perversity."

To return to the *Time* images, they illustrate the dangers lurking behind cyberporn: overexposure, intrusion, surveillance, and the birth of perverse desires. Anxieties over cyberporn exceed the simple worry over the present conditions. In order to understand cyberporn's ramifications, we are told to imagine a catastrophic future of unbearable and uncontrollable contact.[38] This call assumes that catastrophe could be avoided if pornography were simply purged from this medium. By focusing on pornographic images as the source of vulnerability, those hyping cyberporn perpetuate two "competing" visions of the Internet that are really the obverse of each other: the sunny Information Superhighway and the Smut

---

to thirty years in prison for intent to kidnap, in part due to videotapes of his violent and potentially life-threatening sadomasochistic encounters with willing partners (hanging, electrocution, and so on). See Laura Kipnis, *Bound and Gagged: Pornography and the Politics of Fantasy in America* (Durham, NC: Duke University Press, 1999), 3–63.

38.   The obverse of this is Paul Virilio's fear in *Open Sky* that cybersex will lead to uncontrollable masturbation and the end of physical sex; Virilio compares cybersex to AIDS.

Expressway. For those hyping the Information Superhighway, self- or corporate censorship is key and a rudimentary marketplace of ideas is already in place; for those behind the Smut Expressway, government intervention is needed to create an orderly marketplace. Regardless of these differences, both adhere to a notion of electronic interchange that portrays the ideal user/consumer as fully in control—bathed in the soft light of rationality rather than the glare of publicity or the relentless light of surveillance. Both seek to quarantine the good from the bad, the empowering from the intrusive, the rational from the illogical in order to preserve their vision of communication without noise—communication that proceeds in an orderly fashion, with little or no misunderstanding, with no harassment or irrationality.

The excessive accounts of the Internet's intrusiveness also express anxiety over being in public not quieted by marketplace analogies. Online, one is not simply a spectator-citizen-commodity owner. Even when "just viewing" or "lurking," one actively sends and receives data (all spectators are still visible—the degree of their visibility, or more properly their traceability, is the issue). Dreams of vision from afar coexist with the media's relentless drive toward circulation. As I asserted in the introduction, fiber-optic networks threaten to break the glass so that nothing screens the subject from the circulation of images. Instead of only celebrities being caught in the glare of publicity, average citizens find themselves blinded and harassed. Others' words, transported as light—indeed, translated into light and shooting through glass tubes—invade us. And the computer window does not seem to come with dimming controls. Instead, it engages all acts enlightening—all types of light streaming from a window—from the relentless light of surveillance, to the blinding light of harassment, to the artificial light needed for self-contemplation or self-reflection. Rather than marking an end of the Enlightenment in either sense of the word *end*, the Internet asks us rethink enlightenment so that the act of enlightening is not limited to rational discourse or soft light. Fiber-optic networks, then, physically instantiate and thus explode enlightenment.

The Will to Knowledge

This call to "protect children" from the Internet reminds us that "pornography as a regulatory category was invented in response to the perceived menace of the democratization of culture."[39] Art historian Lynda Nead, analyzing the British trial of Penguin Books (the government sought to stop the paperback release of *Lady Chatterley's Lover*), argues "The concern in 1961 was not so much about the content of the novel, as about the constituency of its audience."[40] In this case, the judge asked the jury, "Would you approve of your young sons, young daughters— because girls can read as well as boys—reading this book? Is it a book you would have lying around in your own house? Is it a book that you would even wish your wife or your servants to read?"[41] Concern with pornography coincides with mass media and the spread of literacy. The judge's reference to women and servants seems anachronistic (perhaps even to those readers in 1961, but appropriate given the content of *Lady Chatterley's Lover*), yet it highlights the fact that pornography's dangers are dangers to unequal others. Given America's insistence on viewing itself as a classless democracy in which every person has an equal opportunity to thrive, the unequal group of choice is children—a group whose "proper" sexual growth is vital to the nation, a group legally incapable of consent prior to the age of sixteen.[42]

This link between mass media and pornography also underscores the relationship between pornography and enlightenment, pornography and the will to knowledge. Pornography was not always considered antithetical to enlightenment, although according to Lynn Hunt, it "developed out of the messy, two-way, push and pull between the intention of authors, artists and engravers to test the boundaries of the 'decent' and the aim of the ecclesiastical and secular police to regulate it."[43] In prerevolutionary France, dealers cataloged pornographic texts as "philosophical," indistinguishable from texts we would now consider to be political or

---

39.   Hunt, *Invention*, 12–13.

40.   Nead, *Female Nude*, 91.

41.   Quoted in ibid.

42.   Prisoners, immigrants, suspected terrorists, felons, and arguably homosexuals all do not have full rights, but they are not viewed as key to the nation's future. I owe this insight to Amy Kapczynski.

43.   Hunt, *Invention*, 10.

philosophical.[44] As Robert Darnton has argued, French Enlightenment thinkers viewed these pornographic texts, which blasphemed the church and the king, as enlightening because their blasphemies enabled the public to "throw off" the tutelage of the church and the king—the range of Denis Diderot's writings, from *L'Encyclopedie* to *Les Bijoux Indiscrets*, reveals the extent to which Enlightenment thinking and pornography were intertwined. This "drive for knowledge" seems ingrained within all pornography, from early Renaissance "primers" about female academes to eighteenth-century confessions, from Victorian secret museums to twentieth-century porn films/videos/photographs, although the specificities of this drive differ from medium to medium, historical era to historical era. Regardless of these differences, pornography portrays its protagonists and readers as voyeurs, who gain secret knowledge through their spying or "lurking." Gertrude Koch asserts that "all film pornography is a 'drive for knowledge' that takes place through a voyeurism structured as a cognitive urge."[45] This claim coincides with Darnton's assessment of prerevolutionary print pornography:

If any tendency distinguished this category [popular sex books of prerevolutionary France] as a whole, it was voyeurism. Everywhere in the libertine talks, characters observed one another through keyholes, from behind curtains, and between bushes, while the reader looked over their shoulders. Illustrations completed the effect. In fact, they often showed couples copulating before the

---

44. Pornography would seem a form of blasphemous knowledge. Although most historians argue that pornography became depoliticized when it was separated from other forms of "philosophy" in the nineteenth century, pornography—specifically "deviant" pornography—maintains a political edge given the politicization of "love" and "family" in nineteenth- and twentieth-century England and the United States. Deviant pornography blasphemes the family, Darwinian sexual selection, and heterosexual normativity, just as French prerevolutionary pornography blasphemed Catholicism and the king. The twentieth-century text that best approximates prerevolutionary pornography is *The Starr Report*. In it, Kenneth Starr reveals titillating details of the President Clinton's sex life in order to "dethrone" him.

45. Quoted in Linda Williams, *Hardcore: Power, Pleasure, and the "Frenzy of the Visible"* (Berkeley: University of California Press, 1999), 48.

secret gaze of a narrator, who might be masturbating as if he or she (frequently she) were inviting the reader to do the same. Lascivious putti or shocked prudes frequently looked down on the scene from pictures within the picture. The interplay of illustration and text multiplied the effect of mirrors within mirrors, giving an air of theatricality to the whole business—and … it was often philosophical as well.[46]

The position of the (masturbating) subject who sees, but is not seen, which pornography mimes, is the mythic position of power: the position of the colonizing subject, the guard in the central tower of the panopticon, the cinematic spectator. In many ways, this position of control seems a compensation for the lack of bodily control we experience when watching or reading a pornographic text.

This voyeuristic stance depends on realism: the force of these images stems from their "transparency," from the ways they seem to move beyond representation to reference through our own visceral reaction to them; yet the extreme realist quality of visual and textual pornography is a style that denies being one. As Lynda Nead contends, "There is of course a paradox in this conception of pornography as stylelessness, or of style reduced to the utmost degree. For language—written or visual—to give the reader a sense of stylistic absence demands extreme stylization."[47] Realism enables voyeurism, enables the structure of the will to knowledge, and is itself an Enlightenment mode of narrative. Medium does, however, make a difference. Linda Williams has most forcefully argued in *Hard Core: Power, Pleasure, and the "Frenzy of the Visible"* that "machines of the visible" operate via a principle of maximum visibility: cinema does not simply enhance perverse desires, such as voyeurism, already present in the subject; it produces a new, larger-than-life body that is ideally visible. On display for the viewer, it goes about its business as if it were unaware of being watched. Conversely, cinematic representation provides the

---

46.   Darnton, *Forbidden*, 72–73.

47.   Lynda Nead, "'Above the Pulp-Line': The Cultural Significance of Erotic Art," in *Dirty Looks: Women, Pornography, Power*, eds. Pamela Church Gibson and Roma Gibson, 148.

spectator with a seemingly perfected form of invisibility—an ideal position from which to witness confessions of pleasure.

Pornographic Webcams, examined in more detail in chapter 5, offer their users a similar (delusional) perspective: a penetrating view of their "amateur" models, although their images are certainly less overwhelming than cinematic ones.[48] Webcam sites featuring washroom or changing room cams—which are more often than not faked—coincide with the desire, in *Les Bijoux Indiscrets*, to reveal involuntarily the truth of sex.[49] Regardless, pornography offers an obscene answer to that unanswerable question, What do women want? or more generally, What does the other want? Pornography does not pervert learning but rather reveals the perversion inherent in knowledge acquisition, inherent in "lurking." Importantly, even though online pornography manifests a certain "knowledge-pleasure," it differs in content and form from other types of pornography. These medium-related differences, as Linda Williams has argued, are key. (As I contend in more detail later, rather than simply contributing to what Williams calls "the frenzy of the visible," online pornography leads to a frenzied display of the decline of the visible.) Although all Webcam sites play with voyeurism, they are not simply voyeuristic. The popular IsabellaCam advertises itself as "100% REAL": "Ready to change your mind about online sex? ready 2 experience a real girl who is living

---

48.　Most of these sites are professional, and rather than being "run by" the models themselves, they are managed by men who own several sites. For more on this, see Frederick S. Lane, *Obscene Profits: The Entrepreneurs of Pornography in the Cyber Age* (New York: Routledge, 2000).

49.　Foucault argues that his *History of Sexuality* seeks to "transcribe into history the fable of *Les Bijoux Indiscrets*." According to Foucault,

For many years, we have all been living in the realm of Prince Mangogul: under the spell of an immense curiosity about sex, bent on questioning it, with an insatiable desire to hear it speak and be spoken about, quick to invent all sorts of magical rings that might force it to abandon its discretion. As if it were essential for us to be able to draw from that little piece of ourselves not only pleasure but knowledge … a knowledge of pleasure, a pleasure that comes of knowing pleasure, a knowledge-pleasure. (*History*, 77).

As Julie Levin Russo has argued, the fact that the women in Diderot's novel speak their desire rather than being surveilled marks an important difference (personal correspondence).

out her own explicit sexual fantasies live on cam for all 2 see? had enough with corporate porn and fake orgasms? come here and open your mind 2 the ultimate in virtual sexual experiences where everything u see is 100% REAL"[50] Directly appealing to the viewer's will to knowledge, these "amateur" Webcam sites (both authentic and fake) do not simply generalize or spread voyeurism (the users are *invited* to watch) but rather *mimic* voyeurism in order to create indexicality and authenticity within a seemingly nonindexical medium. As Thomas Levin has observed in his analysis of new cinematic practices (which themselves are inspired by closed-circuit television), the "live" has come to signify the real.[51] This new "realness" incorrectly presumes that "live" material cannot lie, cannot be digitally altered like other online materials. Still, "voyeuristic" images lend the Internet an authenticity it otherwise does not have, and "nonpornographic" cam sites, such as JenniCam, flirted with nudity in order to prove their "realness."

Webcams—especially those that are "interactive"—spread simulated voyeuristic/sadistic pleasure to nonpornographic Web sites, further buttressing the "reality effect" necessary to making fiber-optic communication and computer-generated images seem transparent (whether or not these images are real, they are still generated from code, rather than simply relayed). These interactive Web sites nicely reveal the reason why *agency* has become the word to describe one's ability to act. For instance, during Coco Fusco's live Internet performance piece *Dolores 10–22*—which reenacted the situation of a Mexican worker who, caught with union materials in her purse, was confined in a room for twelve hours until she finally submitted a forced letter of resignation—one active online participant kept insisting that this performance be interactive and pushed for her boss, played by Ricardo Dominguez, to perform sadistic torture as their agent. While others ignored or tried to reason with this individual (most participants were, after all, electronic artists or hactivists), many did turn their

---

50.　⟨http://www.topcams.com/home⟩ (accessed September 1, 2000).

51.　For more on this, see Thomas Y. Levin, "Rhetoric of the Temporal Index: Surveillant Narration and the Cinema of Real Time," in *CTRL [SPACE]: Rhetorics of Surveillance from Bentham to Big Brother*, eds. Thomas Y. Levin et al. (Cambridge: MIT Press, 2002): 578–593.

attention to the ways in which Webcams seem to reduce political witnessing to an act of voyeurism. Fusco herself has condemned the Web for being essentially pornographic, for spreading a sadistic voyeurism. Whether or not the Web is essentially pornographic, certain interactive setups do seem to perpetuate cruelty, pleasure, or both through their conflation of both senses of the word *agency*, through their laboratory of knowledge.

Viewing Webcams and downloading images are part of a general "will to knowledge"—the thrill gained from these activities partly stems from the sexualization of power and resistance itself:

The medical examination, the psychiatric investigation, the pedagogical report, and family controls may have the over-all and apparent objective of saying no to all wayward or unproductive sexualities, but the fact is that they function as mechanisms with a double impetus: pleasure and power. The pleasure that comes of exercising a power that questions, monitors, watches, spies, searches out, palpates, brings to light; and on the other hand, the pleasure that kindles at having to evade this power, flee from it, fool it, or travesty it. The power that lets itself be invaded by the pleasure it is pursuing; and opposite it, power asserting itself in the pleasure of showing off, scandalizing, or resisting. Capture and seduction, confrontation and mutual reinforcement; parents and children, adults and adolescents, educator and students, doctors and patients, the psychiatrist with his hysteric and his perverts, all have played this game continually since the nineteenth century. These attractions, these evasions, these circular incitements have traced around bodies and sexes, not boundaries not to be crossed, but *perpetual spirals of power and pleasure*.[52]

Evading public norms and downloading "deviant" pictures thus may be erotically charged, but not because, as Elmer-Dewitt assumes, these images necessarily correspond to one's "sexuality." Rather, these evasions and travesties—the downloading of images that do not represent the vanilla sexuality that most Americans reportedly enjoy—perpetuate spirals of power and pleasure, spreading sexuality everywhere, making database categories—its basic units of knowledge—sexually charged. Power is therefore experienced as sexuality.

52.   Foucault, *History*, 45.

Internet search engines highlight the spreading of sexuality over al-most every identity (now database) category. If, as Williams alleges, video pornography spawned numerous new genres of pornography such as am-ateur, bondage, and discipline, Internet pornography has expanded the number of categories by several orders of magnitude.[53] The popular por-nography search engine penisbot.com, for instance, lists as its "Straight" categories: Amateur, Anal Sex, Asian, Babes, Black, Celebs, Close Ups, Cum Shots, Ethnic, Group Sex, Hardcore, Interracial, Latin, Lesbians, Masturbation, Megasites, Oral Sex, Porn Stars, Products, Public Nudists, Softcore, Teens, Video, and Webcams. Under its "Men" section, it lists: Amateur, Asian, Bdsm, Bears, Bizarre, Black, Body Builders, Celebs, Eth-nic, Fetish, Hardcore, Hunks, Interracial, Latin, Leather, Megasites, Older Men, Porn Stars, Products, Softcore, Transsexuals, Twinks,

---

53.   Williams, *Hardcore*, 303. Cyberporn transforms pornography by playing with the notion of "numbers," which Williams argues structures "time-based" pornography. Narrative, Williams says, is not pressing in pornography, which like Dziga Vertov's *Man with a Movie Camera*, is structured as a string of "events": the sex scenes are like numbers in a musical, mediating narrative oppositions with sex-ual union. In cyberporn, numbers matter far more numerically. Not only is there an emphasis within porn megasites on the number of porn sites accessible from them (porn sites in turn emphasize the number of pictures available on their site), but the more "narrative" (read time-based) sites treat their objects like numbers, employing the same frame around each "episode" (as do video pornography se-ries). Sites such as 8th Street Latinas, bangbus.com, and milf.com that use "exploi-tation" as their main narrative, feature each "victim" in the same identical frame (often in the same position) on their introductory pages. The accompanying text usually describes the woman and/or the ways in which the owners of the site and the woman met (which is crucial for sites such as bang.bus, which claims to pick up random people with the offer of a lift, have sex with them, and then leave them in the middle of nowhere; on 8th Street Latinas, women are lured with the promise of assistance in obtaining a green card). The sample movies provided all follow the same structure: we see the initial contact, some form of penetration and fellatio, a facial, and then the "victim's" unceremonious dismissal. The user is in on the gag, and this nicely screens the fact that the user too is manipulated and tracked, if not exploited (in truth, these women are probably more in on the gag than the users). Even nonovertly exploitative porn sites that feature goths who strip in order ex-press their freedom and have significant numbers of female members, such as suicidegirls.com, use an unchanging frame for their "girls."

Uniforms, and Video; under "Fetish": Bbw, Bdsm, Bizarre, Blondes, Breasts, Brunettes, Feet & Legs, Fem Dom, Fetish, Fisting, Hairy, Lingerie, Older & Mature, Petite & Midgets, Pregnant, RedHeads, Sex Toys, Shaved, Smoking, Smothering, Spanking, Uniforms, Voyeur & Upskirts, Water Sports, and Zoo Fetish; and under "Other": Adult Games, Bisexuals, Cartoons, Dating, Escorts, Humor, Link Sites, Sex Stories, Shops, and Products. Notably excluded is lesbian as a major category rather than a subset of straight pornography, and female-female sex plus female-male sex is not considered bisexual. There are, of course, lesbian- and bisexual-centered sites, and these omissions, representative of mainstream pornography, are hardly surprising for a site called penisbot that uses the clit counter to track its users. Unlike other media, Internet porn sites allow users to sample readily between its numerous categories (which contain many of the same images). In a video store, one is exposed to lots of different kinds of pornography, but one must rent a video in order to see more images than those on the jacket. On the Internet, one can easily click on these sites to "see" these images, and all these categories are one click away from each other. This explosion in pornographic categories reveals the will to knowledge specific to post-CDA Internet pornography—one linked to knowledge as database and the user as chooser. (The "new" database pornography differs significantly from pre-Web Internet pornography, which was notoriously difficult to index because it was disseminated mainly through newsgroups and BBSs.)

Lev Manovich, in *The Language of New Media*, has argued persuasively that databases, along with navigable space, are *the* two forms of new media. A database is a collection that unlike a traditional collection, "allows one to quickly access, sort, and reorganize millions of records; it can contain different media types, and it assumes multiple indexing of data, since each record besides the data itself contains a number of fields with user-defined values." According to Manovich, "The Internet, which can be thought of as one huge distributed media database, also crystallized the basic condition of the new information society: over-abundance of information of all kinds."[54] This database structure has migrated "back into

---

54. Lev Manovich, *The Language of New Media* (Cambridge: MIT Press, 2001), 214, 35.

culture at large, both literally and conceptually. A library, a museum—in fact, any large collection of cultural data—is replaced by a computer database. At the same time, a computer database becomes a new metaphor that we use to conceptualize individual and collective cultural memory, a collection of documents or objects, and other phenomena and experiences."[55] Examining this "transcoding" of the database, Manovich diagnoses "database complex," an irrational desire to preserve and store everything. Databases challenge traditional understandings of narratives as well as collections: database is the "unmarked term" of the binary opposition between narrative and databases, for databases enable narrative, but narrative does not enable databases. In order to understand this new challenge to narrative, Manovich returns to Dziga Vertov's *Man with a Movie Camera* as a fruitful way to think through the relationship between database and narrative.

Cyberporn sites spin database complex differently, and their redundancy exposes the fact that the *possibility*, rather than the actuality, of overwhelming data provokes desire and panic (Manovich himself remarks that "porno Web sites expose [ ] the logic of the Web at its extreme by constantly reusing the same photographs from other porno Web sites.... Thus, the same data [gives] rise to more indexes than the number of data elements themselves.").[56] As well, the Internet is not a database. As Wolfgang Ernst has noted, the archive has become "metaphorical" in the age of the Internet. "The Internet," he argues, "has no organized memory and no central agency, being defined rather by the circulation of discrete states."[57] Search engines do not search the Internet but rather their own databases, which are produced through "robots," also known as "spiders" or "crawlers," that travel through the Web requesting and storing files (this is why Google can offer cached versions of Web pages). Porn sites and search engines thus offer a false impression of electronic data's

55.   Ibid., 214.

56.   Ibid., 225.

57.   Wolfgang Ernst, "Discontinuities: Does the *Archive* Become Metaphorical in Multi-media Space?" In *New Media, Old Media: A History and Theory Reader*, eds. Wendy Hui Kyong Chun and Thomas Keenan (New York: Routledge, 2005), 105–123.

accessibility and expanse. This impression of accessibility and expanse has driven legislators to create new laws, adding another twist to the intertwinings of power and pleasure coating fiber-optic networks.

Pornocracy

Responding to cyberporn's "dangers," the U.S. legislature enacted two laws, the CDA (1996) and COPA (1998), that restricted minors' access to "unsuitable materials." Leaning on a series of court decisions that have placed the protection of minors above the First Amendment rights of adults, the government cited the need to shield minors as compelling interest.[58] These acts, unlike other laws designed to regulate media content, sought to protect speakers by commercializing pornography and indecency, by creating a soft and fuzzy public sphere in which people literally "buy and sell" ideas.

In the United States, twentieth-century battles over the limits of mass media have centered on pornography and obscenity. Each U.S. Supreme Court decision offers a different relationship between regulatory and disciplinary power, a different way of understanding the relationship between allowable and forbidden speech, while at the same time always giving the Court power to decide the limits of the "speakable."[59] According to *Kovacs*

---

58.   *New York v. Ferber*, 458 U.S. 747m 757 (1982) (quoting *Globe Newspaper Co. v. Superior Court*, 457 U.S. 596, 607 [1982]; and *Sable v. FCC*, 492 U.S. [1989]).

59.   In terms of pornography legislation, in 1957 (*Roth v. the United States*), the Supreme Court decided the First Amendment did not extend to "obscenity," but it also defined obscenity to be materials that are "utterly without redeeming social importance," which to "the average person, applying contemporary local standards, the dominant theme taken as a whole appeals to prurient interests" (354 U.S. [1957]) As Frederick Lane has argued, although the *Roth* decision upheld Samuel Roth's indictment for mailing what we would now consider to be "fairly mild sexual materials," it also enabled the genesis of magazines such as *Playboy* and *Penthouse* as well as the distribution of literary works such as *Lady Chatterley's Lover* and *Memoirs of a Lady of Pleasure* because of the phrases "local standards," "taken as a whole," and "utterly without redeeming social importance" (*Obscene Profits*, 25). This decision was revised by *Miller v. California* in 1973, so that the new three-pronged test for obscenity became (a) whether the average person, applying contemporary community standards, would find that the work, taken as

v. Cooper (1949), "The moving picture, the radio, the newspaper, the handbill, the sound truck and the street corner orator have differing natures, values, abuses and dangers. Each … is a law unto itself."[60] Because broadcast "invades" the home, and because children can hear and see before they can read, broadcast receives the least First Amendment protection, while cable and the telephone receive more.[61] These decisions reveal the intimate relationship between pornography legislation and mass media: without mass commodification, transmission, and production, there would be no pornography legislation (if not pornography itself).[62] These decisions also explain why the debate over Internet regulation was centered on analogies: Was the Internet like broadcast, or was it like the telephone or print? Was the Internet mainly filled with images or text?

a whole, appeals to the prurient interest; (b) whether the work depicts or describes, in a patently offensive way, sexual conduct specifically defined by applicable state law; and (c) whether the work, taken as a whole, lacks serious literary, artistic, political or scientific value. The "Miller test" made more explicit the relation between pornography (as the depiction of sexual acts) and obscenity, and it took "sexual conduct" outside of considerations applied to the work taken as a whole. It also moved from local to community standards, thus implying that something like a "community" with appropriate standards existed. Lastly, its narrowing of "utterly without redeeming social importance" to lacking "serious literary, artistic, political or scientific value" further restricted "free speech." Although Miller is still arguably the text for "print obscenity," each new communications medium required a "new" decision in order to demarcate the "unspeakable," and therefore speakable words/images. For instance, the question of "community standards" becomes a key point of contention in defining online obscenity. The CDA sought to avoid this by simply taking out community standards in its creation transmission clause. For more on the limits of the speakable, see Judith Butler, *Excitable Speech: A Politics of the Performative* (New York: Routledge, 1997).

60.   *Kovacs v. Cooper*, 336 U.S. (1949).

61.   See *Sable v. FCC*, 492 U.S. (1989); *FCC v. Pacifica*, 438 U.S. (1978); *Red Lion Broadcasting Co. v. FCC*, 395 U.S. 367 (1969); *Turner Broadcasting Systems v. FCC*, 114 St. Ct. (1994); and *Miami Herald Publishing Co. v. Tornillo*, 418 U.S. 214 (1974).

62.   For more on this, see Paula Findlen, "Humanism, Politics, and Pornography in Renaissance Italy," in *The Invention of Pornography*, ed. Lynn Hunt, 49–108.

Faced with the privatization of the Internet backbone in 1994–1995, the U.S. government passed the CDA, an act key to understanding the Internet as both a threat to and enabler of democracy. The CDA threatened with fines (up to $100,000), imprisonment (up to two years), or both, "anyone who makes, creates, solicits, and/or initiates the transmission of any communication that is obscene or indecent, knowing that the recipient is under eighteen years of age, regardless of who placed the call." It also threatened to do the same to "anyone who displays, in a manner available to anyone under eighteen years of age, any communication that, in context, depicts or describes, in terms patently offensive as measured by contemporary community standards, sexual or excretory activities or organs, regardless of who placed the call." Lastly, it threatened to fine and/or imprison anyone who knowingly permits any telecommunications facility under this control to be used for such communications. The government would offer safe harbor to those who have, in good faith, taken reasonable, effective, and appropriate actions to restrict or prevent access by minors. Verified credit cards, debit accounts, adult access codes, or adult personal identification numbers—all methods employed by commercial pornography sites in 1996—were named as adequate restrictions.

Although the CDA revised provisions initially aimed at regulating telephony, the crux of the CDA was an analogy between cyberspace and broadcast: the Internet, like broadcast, "invades the home." Further, the Department of Justice in its brief to the Supreme Court argued that the Internet was worse than broadcast: "Because millions of people disseminate information on the Internet without the intervention of editors, network censors, or market disincentives, the indecency problem on the Internet is much more pronounced than it is on broadcast stations."[63] Because of this lack of intervention, while the Internet has "incredible potential as an education and information resource," "that same technology ... allows sexually explicit materials, including 'the worst, most vile, [and]

---

63. U.S. Department of Justice, "C. The Display Provision Is Facially Constitutional, 1.a," in *Department of Justice Brief (Reno v. ACLU), filed with the Supreme Court on January 21, 1997,* ⟨http://www.ciec.org/SC_appeal/970121_DOJ_brief.html⟩ (accessed May 21, 1998).

most perverse pornography,'" to be "only a few click-click-clicks away from any child."[64]

This click-click-click proximity of Net porn compromises the efficacy of zoning laws. As Senator Daniel R. Coats put it during the congressional debate over the CDA, "Perfunctory onscreen warnings which inform minors they are on their honor not to look at this [are] like taking a porn shop and putting it in the bedroom of your children and then saying 'Do not look.'"[65] The government moved toward zoning partly because cyberspace lends itself to questions of spatial segregation, and partly because the CDA leaned on previously upheld zoning laws to prohibit the display of obscene and indecent materials. According to the Department of Justice, "The display provision operates an adult 'cyberzoning' restriction, very much like the adult theater zoning ordinance upheld in Renton and Young."[66] Through this move, the geography of the physical world and cyberspace are correlated; thus, concerns over pornography are "directly analogous to the concerns about crime, reduced property value, and the quality of urban life."[67] Since the porn shop resides in the bedroom rather than on the street, zoning becomes a more pressing and intimate issue.

Zoning regulations, which restrict the display of "indecent" materials to certain commercial zones, combined with credit card verification as a safe harbor, seek to protect access to sexual content by commercializing all sexual content. This effectively moves regulation from the auspices of the government to the market, while at the same time enormously expanding the materials to be regulated (COPA makes this strategy more explicit). The government thus protects free speech by making it no longer free. The CDA, with its safe harbor of credit card–based age verification, effectively forces all obscene or indecent content providers to

---

64.   U.S. Department of Justice, "Summary of Argument," in *Department of Justice Brief*, and "Statement, 2," in *Department of Justice Brief*.

65.   Quoted in U.S. Department of Justice, "Statement, 2."

66.   U.S. Department of Justice, "Summary of Argument, B," in *Department of Justice Brief*.

67.   U.S. Department of Justice, "C. The Display Provision."

become commercial; as the noncommercial plaintiffs such as Stop Prisoner Rape argued, the costs of employing such a system are prohibitive. This "negligence standard," as the legal scholar and activist Amy Kapczynski contends, in contrast to a strict liability standard, enables the relatively free flow of commercial pornography and the discretion to self-regulate.[68] The "problem" the CDA attacks is not commercial pornographers but rather entities that provide pornography—or more properly "indecent materials"—for free. Residing outside market forces, without the pressures of having to sell programming to advertisers and the general public, these entities make the "vilest" materials readily available, seemingly out of the goodness of their own hearts. The Senate justified the forced commercial regulation of "indecency" by maintaining that providers, rather than parents, should shoulder the monetary burden.[69]

For the future of our children, then, the CDA sacrificed the *free* circulation of some ideas. Or to spin it more attractively—as the Department of Justice did in response to the Eastern District Court's decision to grant a preliminary injunction against the enforcement of the CDA—Congress decided that it must stop the free circulation of some obscene ideas in order to ensure the free flow of others, in order to make cyberspace truly public, where public means free from pornography. According to the *Department of Justice Brief*, the inadequate segregation of pornography from the rest of the Internet effectively violated the rights of adults since "the easy accessibility of pornographic material on the Internet was deterring

---

68. In legislation designed to protect minors, there is some tradition of strict liability. For instance, the "reasonable belief" that your wife was over sixteen is not a defense against a statutory rape charge. According to Amy Kapczynski, "This negligence standard embodies a certain kind of Foucauldian regulation—you have to imagine what the 'reasonable person' would do to keep this stuff from kids, and are thus allowed a certain discretion to self-regulate, as opposed to having definitive rules that were imposed by the state" (personal correspondence).

69. According to Senator Charles Grassley, it is not fair for parents to have "the sole responsibility to spend their hard-earned money to ensure that cyberporn does not flood into their homes through their personal computers" (quoted in U.S. Department of Justice "E. There Are No Alternatives That Would Be Equally Effective in Advancing the Government's Interests," in *Department of Justice Brief*).

its use by parents who did not wish to risk exposing their children to such material."[70] Through this argument, the Department of Justice sidestepped the relationship between access and infrastructure/income/education while also appearing to support access.[71]

This reasoning, however, failed to persuade the judiciary of the CDA's constitutionality.[72] In response to the attorney general's argument that the CDA follows precedents set for broadcast regulation, the Supreme Court decided that "the special factors recognized in some of the Court's cases as justifying regulation of the broadcast media—the history of extensive government regulation of broadcasting ... the scarcity of available frequencies at its inception ... are not present in cyberspace. Thus, these cases provide no basis for qualifying the level of First Amendment scrutiny that should be applied to the Internet."[73] Given that cyberspace, unlike broadcast media, receives "full" First Amendment protection, the vagueness of the terms *indecent* and *patently offensive* become crucial: without *FCC v. Pacifica* to rely on (because Pacifica was restricted to broadcast), indecent does not have a judicial history; Congress's definition of patently offensive leaves open the question of whose

---

70.   Department of Justice, "Statement, 2."

71.   The government thus works to ensure that public spaces are legally available to all, without addressing issues of fair access (just as after the battle over civil rights, it ensured that race-based barriers were taken down, but did not address inequalities in income and opportunity in a manner that would guarantee fair access to these public sites).

72.   The zoning argument did, though, win over Justices Sandra Day O'Connor and William Rehnquist. In their *Concurrence*, O'Connor notes, "I write separately to explain why I view the Communications Decency Act of 1996 (CDA) as little more than an attempt by Congress to create 'adult zones' on the Internet. Our precedent indicates that the creation of such zones can be constitutionally sound. Despite the soundness of its purpose, however, portions of the CDA are unconstitutional because they stray from the blueprint our prior cases have developed for constructing a 'zoning law' that passes constitutional muster" *(Concurrence by O'Connor/Rehnquist): Reno v. American Civil Liberties Union et al.*, ⟨http://www.ciec .org/SC_appeal/concurrence.html⟩ (accessed September 19, 1997).

73.   Supreme Court, *Syllabus of Supreme Court Decision in Reno v. ACLU*, ⟨http:// www.ciec.org/SC_appeal/syllabus.html⟩ (accessed September 19, 1997).

community standard is pertinent, and is without the usual clauses about artistic merit and parental support.[74] The vagueness of these words—or the lack of a bright line—causes individuals to steer clear of constitutionally protected speech and deprives the medium of its richness in content.

Thus, according to the Supreme and District Courts' decisions, Congress did not adequately tailor the CDA to the medium. Whereas broadcast is marked by scarcity, pervasiveness, and intrusiveness (thereby enjoying the least First Amendment protection), the Internet is distinguished by plenitude and user participation. Specifically,

four related characteristics of Internet communication have a transcendent importance to our shared holding that the CDA is unconstitutional on its face.... First, the Internet presents very low barriers to entry. Second, these barriers to entry are identical for both speakers and listeners. Third, as a result of these low barriers, astoundingly diverse content is available on the Internet. Fourth, the Internet provides significant access to all who wish to speak in the medium, and even creates a relative parity among speakers.[75]

These four characteristics make the Internet "the most participatory form of mass speech yet developed ... [and thus] deserves the highest protection

---

74.   In response to the government's precedents, the Supreme Court stated that a close look at the precedents relied on by the Government—*Ginsberg v. New York*, 390 U.S. 629; *FCC v. Pacifica Foundation*, 438 U.S. 726; and *Renton v. Playtime Theatres, Inc.*, 475 U.S. 41—raises, rather than relieves, doubts about the CDA's constitutionality. The CDA differs from the various laws and orders upheld in those cases in many ways, including that it does not allow parents to consent to their children's use of restricted materials; is not limited to commercial transactions; fails to provide any definition of "indecent" and omits any requirement that "patently offensive" material lack socially redeeming value; neither limits its broad categorical prohibitions to particular times nor bases them on an evaluation by an agency familiar with the medium's unique characteristics; is punitive; applies to a medium that, unlike radio, receives full First Amendment protection; and, cannot be properly analyzed as a form of time, place, and manner regulation because it is a content based blanket restriction on speech. (*Syllabus*)

75.   District Court for the Eastern District of Pennsylvania "D. 3. The Effect of the CDA and the Novel Characteristics of Internet Communication," in *American Civil Liberties Union v. Reno*, no. 99–1324 (hereafter referred to as *Preliminary Injunction*).

from governmental intrusion."[76] This characterization highlights the act of posting (most surfers do not post to newsgroups, listservs, or the Web) and ignores the nonvolitional "speech" driving Internet protocol. Regardless, it was decided that the government, rather than pornography, intrudes.

Judge Stewart Dalzell, in granting the temporary injunction against the CDA, quotes from Justice Oliver Wendell Holmes's famous dissent: "When men have realized that time has upset many fighting faiths, they may come to believe even more than they believe the very foundations of their own conduct that the ultimate good desired is better reached by free trade in ideas—that the best test of truth is the power of the thought to get itself accepted in the competition of the market."[77] Prior to the Internet, this theory seemed "inconsistent with economic and practical reality." Economic realities have skewed the marketplaces of mass speech in favor of "a few wealthy voices . . . [that] dominate—and to some extent, create—the national debate. . . . Because most people lack the money and time to buy a broadcast station or create a newspaper, they are limited to the role of listeners, i.e., as watchers of television or subscribers to newspapers."[78] To worsen the situation, economic realities have forced competing newspapers to consolidate or leave the marketplace, effectively leaving most Americans with no local competing sources of print media. Lastly, cable has not delivered on its promise to open the realm of television. "Nevertheless, the Supreme Court has resisted governmental efforts to alleviate these market dysfunctions [since] . . . the Supreme Court held that market failure simply could not justify the regulation of print."[79] With the advent of the Internet, however, the judiciary can go on the offensive by simply preserving indecency.

According to Dalzell's decision, the presence of indecency proves the diversity of the medium: "Speech on the Internet can be unfiltered,

---

76. District Court for the Eastern District of Pennsylvania "E. Conclusion," in *Preliminary Injunction*.

77. Quoted in *Abrams v. United States*, 250 U.S. 616, 630 (1919).

78. District Court for the Eastern District of Pennsylvania "D. 4. Diversity and Access on the Internet," in *Preliminary Injunction*.

79. Ibid.

unpolished, and unconventional, even emotionally charged, sexually explicit, and vulgar—in a word, 'indecent' in many communities. But we should expect such speech to occur in a medium in which citizens from all walks of life have a voice. We should also protect the autonomy that such a medium confers to ordinary people as well as media magnates."[80] Thus, diversity of content stands as evidence of the diversity of people, whether or not such economic or, perhaps more specifically, occupational diversity exists. Without indecency, "the Internet would ultimately come to mirror broadcasting and print, with messages tailored to a mainstream society from speakers who could be sure that their message was likely decent in every community in the country."[81] Indecency thus moves from an evil that must be accepted to proof of democracy, to establishing the "much-maligned 'marketplace' theory of First Amendment Jurisprudence."[82] Albeit in very different terms, Dalzell like Foucault sees pornographic resistance or blasphemous knowledge as supporting, rather than destroying, power.

Judge Dalzell is openly outspoken and enthusiastic in his defense of the Internet, but even Justice Paul Stevens ends his decision by celebrating the phenomenal growth of the Internet, declaring that "the interest in encouraging freedom of expression in a democratic society outweighs any theoretical but unproven benefit of censorship." He also argues that the CDA's breadth is "wholly unprecedented. Unlike the regulations upheld in *Ginsberg* and *Pacifica*, the scope of the CDA is not limited to commercial speech or commercial entities. Its open-ended prohibitions embrace all nonprofit entities and individuals posting indecent messages or displaying them on their own computers in the presence of minors."[83] The

---

80.   District Court for the Eastern District of Pennsylvania "D. 5. Protection of Children from Pornography," in *Preliminary Injunction*.

81.   District Court for the Eastern District of Pennsylvania "D. 3. The Effect of the CDA."

82.   District Court for the Eastern District of Pennsylvania "D. 4. Diversity."

83.   Quoted in U.S. Supreme Court, *Supreme Court Opinion (no. 96–511): Reno v. America Civil Liberties Union et al.*, ⟨http://www.ciec.org/SC_appeal/opinion.html⟩ (accessed September 19, 1997).

Supreme Court—which has repeatedly decided in favor of media monopolies and against antitrust laws, effectively reducing consumer choice—thus stands up for individual citizens in a decision, applauded by all telecommunications companies, that completely ignores the larger implications of the Telecommunications Act of 1996 and the privatization of the backbone. This privatization and the policy-based routing it enabled would profoundly change the Internet.

These decisions hinge on user control. Concentrating on the act of searching and surfing, both the Eastern District and Supreme Court agree with the plaintiffs that "although such [sexually explicit] material is widely available, users seldom encounter such content accidentally.... The receipt of information requires a series of affirmative steps more deliberate and directed than merely turning a dial. A child requires some sophistication and some ability to read to retrieve material and thereby to use the Internet unattended."[84] Rather than being passively attacked by images or speech, a child must deliberately choose indecency. The Internet is not like broadcast, but telephony:

In any event, the evidence and our Findings of Fact based thereon show that Internet communication, while unique, is more akin to telephone communication, at issue in Sable, than to broadcasting, at issue in Pacifica, because, as with the telephone, an Internet user must act affirmatively and deliberately to retrieve specific information online. Even though a broad search will, on occasion, retrieve unwanted materials, the user virtually always receives some warning of its content, significantly reducing the element of surprise or "assault" involved in broadcasting. Therefore, it is highly unlikely that a very young child will be randomly "surfing" the Web and come across "indecent" or "patently offensive" material.[85]

Internet pornography—and by extension, its content in general—does not assault the viewer because the user must click and read. Because one

---

84.   Ibid.

85.   Judge Dolores Sloviter, "C. Applicable Standard of Review," in *Preliminary Injunction*.

usually receives textual descriptions before one receives an image, the random retrieval of indecent or pornographic materials is "highly unlikely." In fact, the question of the random retrieval of smut becomes absorbed into the larger problem of imprecise searches since the technology makes no distinction between decent and indecent materials:

Sexually explicit material is created, named, and posted in the same manner as material that is not sexually explicit. It is possible that a search engine can accidentally retrieve material of a sexual nature through an imprecise search, as demonstrated at the hearing. Imprecise searches may also retrieve irrelevant material that is not of a sexual nature. The accidental retrieval of sexually explicit material is one manifestation of the larger phenomenon of irrelevant search results.[86]

By emphasizing "imprecise searches," the judiciary further highlights user control. The "facts" presume that precise searches do not uncover uninvited and extraneous sites. The Internet is not a porn shop in the bedroom but rather a library or mall with secret exits to porn shops that one accidentally finds by looking too far afield (much like the video store with its pornography section visually cordoned off).

The Supreme Court's ruling did not end the legislature's attempts to regulate Internet content. Instead, Congress intensified its efforts to make Internet content commercial through COPA. At face value, COPA would seem to be a more restricted law since it only prosecutes "whoever knowingly and with knowledge of the character of the material, in interstate or foreign commerce by means of the World Wide Web, makes any communication for commercial purposes that is available to any minor and that includes any material that is harmful to minors." This law takes out the display condition and seems to follow the standard Miller test, by requiring that the text, "taken as a whole, lacks serious literary, artistic, political, or scientific value for minors."

COPA may be limited to commercial speech, but by its definition, most speech on the Web is commercial:

---

86. "Findings of Fact," in *Preliminary Injunction*.

(A) Commercial purposes.—A person shall be considered to make a communication for commercial purposes only if such person is engaged in the business of making such communications.

(B) Engaged in the business.—The term "engaged in the business" means that the person who makes a communication, or offers to make a communication, by means of the World Wide Web, that includes any material that is harmful to minors, devotes time, attention, or labor to such activities, as a regular course of such person's trade or business, with the objective of earning a profit as a result of such activities (although it is not necessary that the person make a profit or that the making or offering to make such communications be the person's sole or principal business or source of income). A person may be considered to be engaged in the business of making, by means of the World Wide Web, communications for commercial purposes that include material that is harmful to minors, only if the person knowingly causes the material that is harmful to minors to be posted on the World Wide Web or knowingly solicits such material to be posted on the World Wide Web.

According to this definition, "free" sites—sites that consumers do not pay to access, but that receive money from advertisers—qualify as commercial speech. That is, the government treats as commercial many Web sites that are free or whose print versions are not regulated as commercial speech, such as the *New York Times* (in print, such regulations apply to advertisements, not to the entire periodical because they carry advertisements). Thus, COPA's definition of commercial speech exceeds the current definitions of print commercial speech, which COPA uses as its precedent. As the Court of Appeals for the Third Circuit judge Lowell A. Reed Jr. argues, "Although COPA regulates the commercial content of the Web, it amounts to neither a restriction on commercial advertising, nor a regulation of activity occurring 'in the ordinary commercial context.'" Although he upheld the *Preliminary Injunction* on the grounds that "community standards" are inapplicable in cyberspace, Reed also stated his "firm conviction that developing technology will soon render the 'community standards' challenge moot, thereby making congressional regulation to protect minors from harmful materials on the Web

constitutionally practicable."[87] Reed is probably referring to digital certificates: electronic identification papers that can reveal the age and the location of the user, among other things. Digital certificates, produced in
reaction to the CDA, but also useful to e-business, reveal that passing
legislation—whether or not it is ever enforced—has a profound impact
on the technological and cultural development of the Web, and that the
U.S. government seeks to legislate as "the invisible hand of the market."

These "failed" efforts to commercialize speech, to use commerce as
a means to govern, have helped transform the Internet from a research/
military system to a mass medium/marketplace. Both the CDA and
COPA offered credit card verification as a safe harbor against prosecution,
even though many minors legitimately own credit cards. The year 1996
marked the transition of online porn from "amateur swapping ... to commercial ventures" because many noncommercial Web sites ceased operating or adopted credit card verification out of fear of prosecution (or desire
for money).[88] The government's listing of credit card verification essentially validated commercial porn sites by making porn sites that charged
for access seem responsible rather than greedy (for charging for something
that was freely accessible elsewhere, for information that as the hacker
adage insists, "should be free"). As such, the threat of government regulation gave Web site developers the necessary "reason" to access their
visitors' credit cards and acclimate them to paying for some information.

The impact of the CDA thus reveals the fact that laws, in order to be
effective, do not need to be enforced or constitutional—laws are no longer
only a form of sovereign power. The government, especially in the age of
"small" government, seeks to impact public and private corporations
"indirectly," through measures that respect the market and corporate
self-regulation. Corporate self-regulation, endorsed by some of the plain-

---

87.   U.S. Court of Appeals for the Third Circuit, *American Civil Liberties Union
v. Reno II*, No. 99–1324, 30, 34.

88.   Weber, "The X Files," A1. The threat of legislation has had a profound impact on Web sites. Altern.org, for instance—a large alternative network in
France—closed down in June 2000 after France passed a law making Web-hosting
services responsible for their users' content. The owner, Valentin Lacambre, did
not wait to see if the law would pass through the French courts.

tiffs, has had a "chilling" effect on Internet speech. For instance, AOL decided in 2003 to reject all e-mail coming from DSL servers in order to reduce spam, and regularly ejects users who do not follow AOL's etiquette
rules. Most of the larger news Web sites carefully filter their content. And
all commercial media organizations constrain content in order to boost
ratings or click throughs.

Crucially, independent noncommercial pornographic or erotic sites
still thrive, using the "click here if you are eighteen to enter the portal"
in order to remain legal—although the Department of Justice in 2004
seemed likely to launch an offensive against cyberporn. Doubtless, small
noncommercial sites exploring nonnormative sexuality would have been
the department's first targets. But given increasing corporate self-
regulation and changes to the fundamental structure of the Internet, what
do individuals now do? Do they, based on the so-called privacy of the
Internet, engage in public acts of nonregulation? And how do acts of nonregulation and the freedom that stems from them relate to questions of
agency?

In Public

The public is the experience, if we can call it that, of the interruption or the
intrusion of all that is radically irreducible to the order of the individual
human subject, the unavoidable entrance of alterity into the everyday life of
the 'one' who would be human.

—Thomas Keenan, "Windows"

The CDA court decisions privilege agency over contact, empowerment
over disruptions, text over images. The Supreme Court's description of
the Web summarizes this conviction nicely:

The Web is thus comparable, from the readers' viewpoint, to both a vast
library including millions of readily available and indexed publications and a
sprawling mall offering goods and services.

From the publisher's point of view, it constitutes a vast platform from which
to address and hear from a world-wide audience of millions of readers,

viewers, researchers, and buyers. Any person or organization with a computer connected to the Internet can "publish" information.[89]

According to the Supreme Court, all users—whether readers, publishers, or both—deliberately act. They read, consume, publish, research, address, listen, or view. They may accidentally retrieve the wrong information and they may, through slips of the keyboard, expose their gender, race, age, and/or physical fitness, but in general, they control what information they receive and send. This deliberateness stems from the textual nature of online communication. Literacy proves a thinking subject. Textual exchange guarantees fair exchange. The Internet, by resuscitating and expanding "print" publishing, restores eighteenth-century optimism.

This conclusion relies on a dangerously naive understanding of language—one that rivals "they wouldn't print it if it wasn't true." It erases the constant involuntary data exchange crucial to any user-controlled exchange of human-readable information, and disastrous to any analogy between print and the Internet. It also assumes an intimate and immediate relation between the written word and the mind, bypassing the unconscious and the ways in which language is beyond the individual. Further, it perpetuates an extremely safe notion of contact between readers and publishers: users do not interrupt each other, stalk each other, or really engage each other at all. Instead, they offer their statements, wait for replies, and perhaps reply back again in an orderly fashion. It assumes that texts can be reduced to ideas, and that people merely consume ideas. Lastly, it assumes that users are always the authors of texts and never their objects: again, the major objection against online pornography was not that it objectified women, as MacKinnon would have it, but rather that it assaulted its viewers. It considers pornographic—and indeed all electronic—intrusion accidental.

This decision also reveals popular belief in the "danger" of images. As mentioned previously, pornographic images are dangerous because they usurp their referent, unless the issue is child pornography—then, the danger stems from their indexicality. U.S. child pornography laws regulate

---

89.   U.S. Supreme Court, *Supreme Court Opinion*.

image-based, and not text-based, pornography. The 1977 Sexual Exploitation of Children Act, the first U.S. law to outlaw the production, sale, circulation, and receipt of child pornography, stated that image-based child
pornography was a "form of sexual abuse which can result in physical or
psychological harm, or both, to the children involved."[90] It also stated
that children were especially vulnerable to these images, for seeing them
could make unwilling victims willing. Although pornographic images do
"move" their viewer, like all images, they are read, and reading predates
writing. As Laura Kipnis puts it, "Pornography grabs us and doesn't let
go. Whether you're revolted or enticed, shocked or titillated, these are
flip sides of the same response: an intense, visceral engagement with what
pornography has to say. And pornography has quite a lot to say.... It's not
just friction and naked bodies.... It has meaning, it has ideas."[91] Kipnis's
insistence on pornography as having meaning is missing in all analyses of
pornography around the CDA and COPA. This insistence on pornography as having meaning also enables a discussion of different kinds of pornography. Just as all books or films are not the same, all pornography is
not the same.

In addition, the Supreme Court's conclusion that the accidental retrieval of pornography results from "imprecise searches" drastically simplifies language. Although adding qualifiers, in proper Boolean fashion,
usually pares down the number of unwanted sites, the unexpected, the
antithetical, and the pornographic do not only emerge when a search is
imprecise. For one, those producing and consuming information are not
cooperating together. Metatags—the tags that determine the site's keywords for which search engines scan—expose this noncooperation (for
instance, Coca-Cola's metatag at one point contained "Pepsi"), as do
pornographic sites that take advantage of typos, such as the porn site
whitehouse.com (versus whitehouse.gov). Marketers, at least, have not dis-
counted the importance of slips of the keyboard, of serendipity; they have
reinserted serendipitous "shopping" by taking advantage of various cracks
in the subject's conscious control. The government, in filing COPA, also

90. "Congressional Findings," "Notes on Sec. 2251," *United States Code.*

91. Laura Kipnis, *Bound and Gagged*, 161.

showed the inadequacies of the imprecise search argument through searches on "toys" and "girls" that produced pornographic sites (also inadvertently complicating simple notions of pedophilia by revealing the widespread sexualization of childhood). As well, a search on "Asian + woman" on Google in 2004 produces more pornographic sites within the first ten hits than one using "pornography."

Exploring what the Supreme Court renders accidental and what high-speed telecommunications networks have made metaphoric—such as archive and vision—reveals the differences the Internet makes. Internet pornography calls into question visual knowledge. Using cinematic pornography as their basis, critics assume that pornography has an all-engrossing visual impact. Fredric Jameson, for instance, asserts in *Signatures of the Visible* that pornographic films are "only the potentiation of films in general, which ask us to stare at the world as though it were a naked body." To Jameson, "the visual is *essentially* pornographic, which is to say that it has its end in rapt, mindless fascination; thinking about its attributes becomes an adjunct of that.... [A]ll the fights about power and desire have to take place here, between the mastery of the gaze and the illimitable richness of the visual object; it is ironic that the highest stage of civilization (thus far) has transformed human nature into this single protean sense."[92] This understanding of the visual as essentially porno-graphic, Jameson admits, stems from cinema and is perhaps not applicable to other media. Regardless, by discussing the Internet within the rubric of pornography, the 1995–1997 debates sought to understand—if not create—the Internet as fascinating through a fundamentally visual para-digm. Fiber-optic networks, however, both enable and frustrate this all-pervasive visuality: visuality, the camera, and the gaze are *effects*, often deliberately employed to make "jacking in" sexy. Although Internet por-nography is visual, its invisible workings are more significant and its visual impact less than that of cinematic pornography.

Pornographic sites notoriously rewrite the basic functions of Web browsers, revealing the ways in which "user choice" is a software con-struction. By rewriting the "back" button, an easy and readily available

---

92.   Fredric Jameson, *Signatures of the Visible* (New York: Routledge, 1992), 1.

javascript, these sites push the user onto another Web site, precisely when s/he wishes to leave. By opening another window when the user seeks to close it, another easy and readily available javascript, they box in the user. The user usually gets stuck in a Web ring and is forwarded from one member site to another, which if the user follows for any length of time, belies notions of endless pornography/information. Porn sites were the first to use the now-standard pop-up window to push images at viewers. These tactics often create panic, since the user has lost "control" over his/her browser. Users also panic when they receive pornographic e-mail messages after visiting certain sites (taking advantage of this panic, porn sites now feature pop-ups by "security companies" that warn you of the porn on your hard drive. By listing the contents of your C drive, they make you believe (mistakenly) that everyone can access your entire hard drive). During the heyday of Netscape 3.0, porn sites used javascripts that culled a person's e-mail address. Although e-mail address capture is more difficult, even the most nonintrusive-seeming sites, such as penisbot.com, collect statistics about user-usage (for example, what site the user last visited or whether or not penisbot is bookmarked). Stileproject.com keeps track of which links have been clicked.

Porn sites take advantage of the many default variables provided by the hypertext protocol and use the latest "trapping" javascripts, while also offering content that reifies users' control. Taking advantage of "live" technology, they offer you models who respond to your commands, who interact with you in the manner that the Supreme Court and the sadistic member of the *Dolores 10–22* chat understand interactivity—your mouse click does seem an affirmative action. They offer you "tours" and give you samples based on your preferences. They enable you to keep a window open for hours, so that while you work, these images patiently wait for you. Online pornography seems less pornographic—less fascinating, less demanding. Pornographic Web sites reveal the tension between and synthesis of individualization and mass interest in their many intro sites, which pick up on porn keywords such as "Oriental" and then push you into a larger site, in which Oriental may or may not be a category. They also pick up on the fantasy of amateur knowledge, of "do-it-yourself" Webbing, through sites supposedly produced by entrepreneurial women models. Thus, the content and the structure of pornography sites expose the tension between freedom and control that underlies the Internet as a new mass medium: on the one hand, it enables greater freedom of expression; on the other hand, it facilitates greater control. Porn site models are amateurs liberating their sexuality or dupes you control. Porn sites enable you to investigate your sexuality without fear of exposure or they track your every move. This opposition of control-freedom erases the constitutive vulnerability that enables communications. It is not either subject or object but both (metaphorically) at once. Publicity stems from the breach between seeing and being seen, between representing and being represented. Publicity is an enabling violence—but not all publicity is the same. The key is to rethink time and space—and language—in order to intervene in this public and to understand how this public intervenes in us, in order to understand how the Internet both perpetuates and alters publicity.

The dangers described by the pro-CDA forces are real: there exists information on the World Wide Web that can play a role in serious tragedies such as the Columbine shootings. Yet democracy has always been about dangerous freedoms, to which the many revolutions to date testify. This is not to say that one must take a libertarian view; this is to say that these "dangers" can also be the most fruitful products of the Internet, that the disruption the Internet brings about can be utilized to formulate a more rigorous understanding of democracy. The key is to refuse hasty leaps between speech and "minds," and between diversity of content and diversity of people.

In short, the Internet is public *because* it allows individuals to speak in a space that is fundamentally indeterminate and pornographic, if we understand pornography to be as Judith Butler argues, "precisely what circulates without our consent, but not for that reason against it."[93] As Keenan remarks,

The public—in which we encounter what we are not—belongs by rights to others, and to no one in particular. (That it can in fact belong to specific individuals or corporations is another question, to which we will return.) Publicity tears us from our selves, exposes us to and involves us with others, denies us the security of that window behind which we might install ourselves to gaze.

93.   Butler, *Excitable Speech*, 77.

And it does this "prior to" the empirical encounter between constituted subjects; publicity does not befall what is properly private, contaminating or opening up an otherwise sealed interiority. Rather, what we call interiority is itself the mark or the trace of this breach, of a violence that in turn makes possible the violence or the love we experience as intersubjectivity. We would have no relation to others, no terror and no peace, certainly no politics, without this (de)constitutive interruption.[94]

In this sense, we are the child—vulnerable to pornography and not yet a discrete private individual. And this position can be terrifying, yet without this we could have no democracy. This chapter, through an examination of cyberporn, has outlined the necessity of this position, the necessity to deal with questions of democracy in terms of vulnerability and fear. Resisting this vulnerability leads to the twinning of control and freedom—a twinning that depends on the conflation of information with knowledge and democracy with security.

We are now facing a turn in what Claude Lefort called the "democratic adventure," and these questions are pressing precisely because it is too easy to accept the Internet as the great equalizer: diversity of content easily becomes an excuse to ignore questions of access; the Internet as the second coming of the bourgeois public sphere easily closes questions of publicity. By questioning the position of the consumer—and its counterpart, the user—we can begin to expose the objectification and virtualization of others that underlie this myth of supreme agency, and begin to understand how the Internet can enable something like democracy. By examining the privatization of language, we can begin to understand the ways in which power and knowledge are changing.

---

94. Thomas Keenan, "Windows: Of Vulnerability," in *The Phantom Public Sphere*, ed. Bruce Robbins (Minneapolis: University of Minnesota Press, 1997), 133–134.

SCENES OF EMPOWERMENT

> One of the wonderful things about the information highway is that
> virtual equity is far easier to achieve than real-world equity.... We
> are all created equal in the virtual world and we can use this equality
> to help address some of the sociological problems that society has yet
> to solve in the physical world.
>
> —*Bill Gates*, The Road Ahead

Race was, and still is, central to conceiving cyberspace as a utopian commercial space. More precisely, conceiving race as skin-deep has been crucial to conceiving technology as screen deep.

Cashing in on mainstream desires for a quick and painless fix to the color line, promoters and visionaries sold the Internet as dissolving the "race problem." Unapologetic capitalists such as Microsoft's Bill Gates and dreamy Californian "homesteaders" such as Howard Rheingold argued that text-based and/or asynchronous image-based communications cemented the blinders on Justice's face (in Gates's case, through a conflation of equity and equality). In stark contrast to actual user demographics, turn-of-the-century advertisements such as MCI's "Anthem" and Cisco Systems's "Empowering the Internet Generation" series featured variously "raced" humans extolling the virtues of global telecommunications networks, which—they informed us—they were already using. According to these and many more advertisements, news reports, and advertisements masquerading as news reports, the future had arrived, and with it, technologically produced social justice. These promotions thus represented as empowering one of the most invasive and insecure forms of communications created to date. Glorifying the power of the mouse click, they transformed the Internet from a pornographic badlands to a user-controlled utopia. As stated earlier, however, rather than simply enabling more people to exercise what Walter Benjamin once called their "legitimate claim to be reproduced," the Internet also circulates their representations without their consent or knowledge.[1] Invisibly, the Internet turns every spectator into a spectacle, and an enormous and unending amount of energy, money, and cultural and computer programming is needed to sustain the Internet as an agency-enhancing marketplace of ideas and commodities.

In this chapter, I examine how corporations sought to blind users to their own constitutive vulnerability—the facts that in order to use, one is used, and that one's online interactions are fundamentally open—by conflating racial and technological empowerment, color- and technology-blindness. This double blindness screened issues of power and discrimination, and transformed the Internet from a U.S. military- and academic research–based "network of networks" to an extraspatial consumerist international. These scenes of empowerment have driven access-based definitions of and solutions to the "digital divide," and have helped make race simultaneously a consumer and pornographic category. This erasure and consumption of race, however, does not make the Internet irrevocably racist; but to fight this trend, antiracist uses of the Internet make race both visible and difficult to consume. They erode the distance between spectator and spectacle sustained by the mainly televisual and literary separation of "users" from "raced others," and attack narratives of "technological empowerment" by refusing to celebrate "ethnic" self-representations as unmediated "amateur" truths. I thus conclude with projects produced by the U.K.- and Jamaica-based digital collective Mongrel—projects that emphasize electronic duplicity in order to expose the racism underlying dreams of a "color-blind" Internet. These works keep open the Internet's promise of democratization, and explore the ramifications and possibilities of vulnerability and connectivity rather than superagency.

1.   Walter Benjamin, "The Work of Art in the Age of Mechanical Reproduction," in *Illuminations: Essays and Reflections*, trans. Harry Zohn (New York: Schocken Books, 1968), 232.

The Race for Users

Sometimes a person doesn't want to seek the dignity of an always-already-violated body, and wants to cast hers off, either for nothingness, or in a trade for some other better model.

—Lauren Berlant, "National Brands/National Body"

MCI's 1997 aptly named "Anthem" commercial epitomized promotions of the Internet as a "medium of minds." This campaign became so influential that almost all Internet-related advertisements in the United States displayed some raced or differently marked flesh (until the postmillennium dot-bomb crisis—then corporations turned to white male images and voice-overs in order to signal stability and experience). "Anthem" features variously raced, gendered, aged, and physically challenged persons chanting, in succession and in concert,

People can communicate mind-to-mind.
There is no race.
There are no genders.
There is no age,
No age.
There are no infirmities.
There are only minds,
Only minds.
Utopia?
No.
No.
The Internet,
Where minds, doors and lives open up.
Is this a great time, or what,
Is this a great time, or what?"[2]

---

2.   This commercial aired throughout 1997. In this chapter, I analyze the shorter version of the commercial. The longer version contains two extra characters: an "ethnic" white male who speaks with a heavy Eastern European accent, and a white male boy in a wheelchair.

Text messages, such as "MCI has the fastest Internet network," appear on a computer screen that punctuates the stream of bodies/body parts, while an upbeat sound track provides continuity.

By picturing electronic text as enabling racial—and indeed, gender and age—passing, telecommunications companies counter arguments of online communications' inferiority to face-to-face ones. They spin what could be considered drawbacks to empowering communications—potential deception and unverifiability—into features enabling "free" and agency-enhancing communications. It is not that someone could be lying to you, or that you cannot be sure who someone is or what they are sending you, but that now *you* can transcend the physical limitations of your *own* body. This positions viewers/would-be users as speakers, rather than listeners, screening the facts that most people "lurk" rather than post and that lurkers "speak" nonvolitionally.[3] By featuring "others" who directly address the audience, this commercial also manipulates the empowerment that supposedly stems from speaking for oneself. In this commercial, as well as Cisco Systems's "From the Mouths of Babes" series (in which young people of color from around the world offer statistics of the Internet's phenomenal growth) and Etrade.com's series (in which a twenty-something woman of color informs the audience that she's "not relying on the government"), pseudosubalterns speak corporate truths.

Significantly, this rewriting of the Internet as emancipatory, as "freeing" oneself from one's body, also naturalizes racism. The logic framing MCI's commercial reduces to what *they* can't see, can't hurt *you*. Since race, gender, age, and infirmities are only skin-deep (or so this logic goes), moving to a text-based medium makes them—and thus the discrimination that *stems from them*—disappear. Although "no race" rather than "no racism" leaves open the possibility of racism without physical markers of "race," this formulation effectively conceals individual and institutional responsibility for discrimination, positing discrimination as a problem that the discriminated must solve. The message is not even "do not discrimi-

---

3.   For example, most people on e-mail lists only read messages from the list—that is lurk—rather than send or post messages. Even if one does not post, however, one still responds by sending the originator a confirmation of receipt at the transport-layer level.

———

nate." It is "get online if you want to avoid being discriminated against." For those always already marked, the Internet supposedly relieves them of *their problem*, of *their flesh* that races, genders, ages, and handicaps them, of *their body* from which they usually cannot escape. Ineffaceable difference, rather than discrimination, engenders oppression, which the discriminated, rather than the discriminators, must alleviate.

However framed, this offer to abandon or trade-in one's always already violated body is tantalizing, and "Anthem" surprisingly supports critics of formal equality (albeit in an attempt to sell the Internet as finally making formal equality equal equality).[4] "Anthem" highlights what many U.S. citizens have been unwilling to admit: namely, that amending the Constitution to include within "the people" those initially excluded has not been enough. Liberty has not guaranteed Freedom. It is not enough that, as Jürgen Habermas notes, "the *status liberatis*, the *status civitatis*, and the *status familiae* gave way to the one *status naturalis*, now ascribed generally to all legal subjects—thus corresponding to the fundamental parity among owners of commodities in the market and among educated individuals in the public sphere." It is not enough that, "however exclusive the public might be in any given instance," adds Habermas, "it could never close itself off entirely and become consolidated as a clique; for it always understood and found itself immersed within a more inclusive public of all private people, persons who—insofar as they were propertied and educated—as readers, listeners, and spectators could avail themselves via

---

4.   For critiques of formal equality, see Bruce Robbins, ed., *The Phantom Public Sphere* (Minneapolis: University of Minnesota Press, 1997), especially Nancy Fraser, "Rethinking the Public Sphere: A Contribution to the Critique of Actually Existing Democracy," Michael Warner, "The Mass Public and the Mass Subject," and Lauren Berlant, "National Brands/National Body: *Imitation of Life*"; Kimberlé Crenshaw, "Color Blindness, History, and the Law," in *The House That Race Built: Black Americans, U.S. Terrain*, ed. Wahneema Lubiano (New York: Pantheon, 1997), 280–288; Patricia J. Williams, *The Alchemy of Race and Rights: Diary of a Law Professor* (Cambridge: Harvard University Press, 1991); Bruce Simon, "White-Blindness," in *The Social Construction of Race and Ethnicity in the United States*, eds. Joan Ferrante and Prince Brown Jr. (New York: Longman, 1998), 496–502; and Ben Bagdikian, *The Media Monopoly*, 5th ed. (Boston: Beacon Press, 1997).

the market of the objects that were subject to discussion."[5] It is not enough that there existed a fundamental parity among owners of commodities in the public sphere, or that the public could not consolidate itself into a clique, because the fundament has never been laid. Inequalities in status have never been adequately bracketed.[6] For the "marketplace of ideas" to work, as Michael Warner has observed, "the validity of what you say in public bears a negative relation to your person. What you say [carries] force not because of who you are but despite who you are." MCI's relentless focus on these people's bodies—or more precisely, their body parts—reveals that "the humiliating positivity of the particular" in real life (or more accurately, real life as portrayed by MCI on television) negates this principle of negativity.[7]

MCI's televisual representation of these raced others reduces these actors to mere markers of difference and displays them for "our" benefit. The power behind "no race, no genders, no age, no infirmities" stems from these raced, gendered, aged, and infirm persons. This positive relation to their bodies, rather than interfering with their speech, grounds it, and grounds "our" assumption that of course *these* people would be happy to be on the Internet. Their physical particularities make these figures generic and interchangeable—it is important that marked persons speak,

---

5.   Jürgen Habermas, *The Structural Transformation of the Public Sphere: An Inquiry into the Category of Bourgeois Society*, trans. Thomas Burger (Cambridge: MIT Press, 1991), 37.

6.   For more on the inadequacies of such attempted bracketing, see Patricia J. Williams, *Alchemy*, 15–43; Warner, "Mass Public"; Berlant, "National Brands"; and Fraser, "Rethinking." In particular, Fraser argues that

this public sphere was to be an arena in which interlocutors would set aside such characteristics as differences in birth and fortune and speak to one another as if they were social and economic peers. The operative phrase here is "as if." In fact, the social inequalities among the interlocutors were not eliminated, only bracketed.... But were they effectively bracketed? The revisionist historiography suggests they were not. Rather, discursive interaction within the bourgeois public sphere was governed by protocols of style and decorum that were themselves correlates and markers of status inequality. These functioned informally to marginalize women and members of the plebian classes and to prevent them from participating as peers. ("Rethinking," 10).

7.   Warner, "Mass Public," 239.

but any persons would suffice, which is why this commercial contains no celebrities. Only the emancipating power of the Internet (specifically MCI's Internet) explains their "surpassing" the (equally) debilitating effects of race, gender, age, and physical infirmities. Shot this way, they make the seamless corporation an attractive alternative to their generically marked bodies. Thus, although this commercial seems to be directed at empowering those "unequal others," it cuts and brands its spokespersons in order to incorporate them. The Internet as a race-free utopia (and subsequently, the user as superagent) relies on, perpetuates, and solidifies the very stereotypes it claims to erase; according to MCI, virtual fluidity comes at the cost of real-life rigidity.

This objectification, this reduction of persons to flesh, follows in the tradition of *pornotroping*. Hortense Spillers, in "Mama's Baby, Papa's Maybe," employs this term to describe the rhetorical uses of the captive body and the continuing "signifying property *plus*" attributed to marked black bodies. Most simply, pornotroping reduces a person to flesh—to a sensuous thing embodying sheer powerlessness—and then displays this flesh to incorporate the viewing subject/body. To bring out the differences between the captive and the "people," Spillers distinguishes between body and flesh, culture and cultural vestibularity. In order for others to become cultured, they pass through flesh (as through a vestibule): "Before the 'body' there is the 'flesh,' that zero degree of social conceptualization that does not escape concealment under the brush of discourse, or the reflexes of iconography."[8] Whereas the body in public is shielded by private protections, flesh is outside the prophylaxis offered by the Fourth and Fourteenth Amendments to the U.S. Constitution. To be flesh is to be open to fissures, scars, and other markings. Flesh is de-gendered. Importantly, bodies cannot emerge as bodies without flesh: there can be no culture or whole without first a vestibule to take the brunt of invasive contact. It is not simply, then, that some have had access to disembodiment and others have not but rather that some have never had a body—in the sense of an

---

8.   Hortense Spillers, "Mama's Baby, Papa's Maybe," in *Within the Circle: An Anthology of African American Literary Criticism from the Harlem Renaissance to the Present*, ed. Angelyn Mitchell (Durham, NC: Duke University Press, 1994), 458, 457.

integrated whole whose skin is seamless and unmarked—from which to abstract in the first place. The fiction of the disembodied citizen depends on this other disembodiment.

MCI's commercial clearly does not perform the extreme violence enacted on the captive slave as discussed by Spillers, but its televisual cuts update what Saidiya Hartman has called "scenes of subjection," which themselves follow from pornotroping. Scenes of subjection, from brutal scenes of whippings to happy portrayals of singing slaves, Hartman argues, express the brutality of slavery in "the forms of subjectivity and circumscribed humanity imputed to the enslaved." Analyzing graphic portrayals of atrocious beatings, she contends that the brutality contained within them was second only to the abolitionist demand that the suffering of slavery "be materialized and evidenced by the display of the tortured body."[9] These scenes produce a form of empathy that obliterates difference: self replaces other, the white self imagines itself the black beaten slave. The other's degradation thus becomes an opportunity for self-reflection, not an event to which one witnesses and testifies. According to Hartman, the forced "happy scenes of slavery" similarly envision blacks "fundamentally as vehicles for white enjoyment," and together these brutal and happy scenes underlie the doctrine of equal but separate.[10] MCI's commercial similarly insists on the inadequacy of these raced, gendered, and aged others and displays them for the pleasure of the television viewer/consumer; but MCI's commercial displays this logic of unequal subjectivity with a twist: these others are happy with their inequality in real life because of their virtual equality elsewhere. The Internet becomes "separate but equal." Through this display of flesh—flesh that the Internet supposedly makes irrelevant—the *televisual fantasy* of the user as superagent emerges.

"Anthem" begins with an upbeat sound track and an old man opening a laptop (figure 3.1). His clothing and the wrinkles on his cutoff face and hands signify his age. This camera angle denies him a "window to the soul" or any facial features that would distinguish him as an individual.

9.   Saidiya Hartman, *Scenes of Subjection: Terror, Slavery, and Self-Making in Nineteenth-Century America* (New York: Oxford University Press, 1997), 6, 4.

10.   Ibid., 23.

| **Figure 3.1** |
Shot from "Anthem"

As a prelude to the message of the commercial (the Internet can offer protection that his skin cannot), however, the computer dominates this shot, protecting his midsection from our view. Next, the commercial cuts to his hands typing on the computer: the computer, arguably, is an extension of this man or a body part (figure 3.2). At this point, the first words dub over the sound track and a little girl says, "People can," which seems to compromise further the physical integrity of this man by separating his body and his voice. The little girl's cutoff face, in color, appears next as she says, "Communicate" (figure 3.3). The commercial then zooms to the old man's face as they together say, "Mind-to-mind," reintegrating this man's body and voice while at the same time melding them together. The contrast between black and white and color emphasizes the differences between the old man and the young girl—differences supposedly bridged by the extracorporeal merging of their voices and, by extension, their minds. The old man's face is not shown until his voice joins with the colorful little girl's, and even then it is cutoff and shown from the perspective of the computer screen looking up at him (figure 3.4). Like the little girl, we do not have immediate physical access to this man, and this double screening (the television and computer screens) would seem to

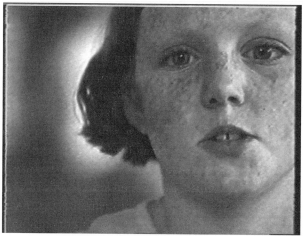

| **Figure 3.2** |
Shot from "Anthem"

| **Figure 3.3** |
Shot from "Anthem"

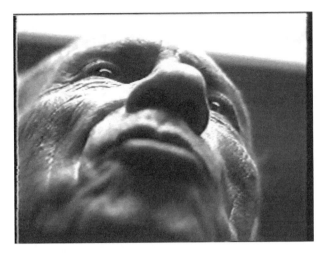

| Figure 3.4 |
Shot from "Anthem"

imply a greater distance between the viewer and the old man, but the extreme close-ups and the telepathic communications between the girl and the old man present another reading: through the Internet, distances shrink and we become closer to each other.

Lest things get too close for comfort, we look down on a black male placed before a window in the next shot (figure 3.5). Here, three windows are in play—the "real" window, the television screen, and the computer screen—and each window signals a different kind of empowerment. The real window overlooking skyscrapers indicates that he has made it to the top, or near the top, of a corporation. Rather than surveying the landscape he has conquered, though, he first gazes into the computer screen (implying that the physical is less important than the virtual) and then looks up at the television viewer. Thus, the television screen is privileged above these windows, and this perspective separates those *others* who look *at* the computer screen and those of "us" (television viewers) who look *through* it: he is analogous to "us," but not us. In fact, he looks up at "us," reassuring "us" that "equality" will not diminish "our" privilege. Through this shot and the concluding ones, in which the initially slightly serious speakers start smiling, MCI tames the already tame defiance of "there is

| Figure 3.5 |
Shot from "Anthem"

no race" by portraying these "persons" as being appeased by Internet-only empowerment. This neatly resolves the contradiction encapsulated by Lauren Berlant: "The nation holds out a promise of emancipation and a pornographic culture both."[11] MCI's answer is: Let your body be placed within a pornographic frame (that is, be objectified and displayed) in real life, but resist objectification (be emancipated) by becoming text online. Technology sutures intractable contradictions and antagonisms.

According to Michael Warner, the viewer's particularity does not prevent its identifying with the "mass viewer" hailed by these advertisements: the peering, coaxing, speaking, and/or smiling faces position "us" as mass subjects, to which they, as particular marked subjects, appeal. Yet the viewer's particularity splits it into the first and third person. "It is at the very moment of recognizing ourselves as the mass subject," he argues, "that we also recognize ourselves as minority subjects. As participants in the mass subject, we are the 'we' that can describe our particular affiliations of class, gender, sexual orientation, race or subculture only as

---

11.   Berlant, "National Brands," 203.

'they.'"[12] Hence, a viewer could be a black male, yet while passing/ addressed as a mass subject, he is separated from *this* black male, whose race marks him (and thus by extension the black viewer) as a subculture: passing produces both the minority subject and the masterful unremarkable one. This commercial, however, also relies on mass subjects misrecognizing themselves as minoritized. Like cyberpunk heroes despairing of their flesh, they identify with these others, or more properly with their supposed desire to rid themselves of this flesh (again, this desire is "naturalized" through a discriminatory logic: of course these people would be happy to be on the Internet). This commercial therefore combines this desire with the viewers' unmarked mobile perspective in order to empower the television viewer, not "raced others." This rerouting of "empowerment" supports and complicates Lisa Nakamura's claim that "networking ads that promise the viewer control and mastery over technology and communications discursively and visually link this power to a vision of the other which, in contrast to the mobile and networked tourist/ user, isn't going anywhere. The continued presence of stable signifiers of otherness in telecommunications advertising guarantees the Western subject that his position, wherever he may choose to go today, remains privileged."[13] It is not only that these others aren't going anywhere but also that no matter where they go, they will always appeal to and establish "us" as mass subjects; they will always invite us to join their heterotopia. This promise of mobility compensates for the viewer's actual lack of mobility and resuscitates the mastery of what Nakamura calls a "Western subject."

Not accidentally, there is only one image of an able-bodied, non-"ethnic" white man (figures 3.6 and 3.7). The picture of domesticity, he flashes on the screen as he hovers over his small daughter, on whom the camera focuses. As he types on the computer, his small daughter watches, and later in the commercial ("Where minds, doors and lives open up"), his daughter takes over the keyboard and we see only his hands surrounding her. In all these shots, this white male—unlike every other figure in this commercial—ignores the audience. The promise of the Internet to

---

12. Warner, "Mass Public," 243.

13. Lisa Nakamura, *Cybertypes: Race, Ethnicity, and Identity on the Internet* (New York: Routledge, 2003), 90.

| Figure 3.6 |
Shot from "Anthem"

| Figure 3.7 |
Shot from "Anthem"

disembody would seem outside his concern, or more precisely, not something he needs to celebrate or acknowledge. Rather than waste his time speaking about the Internet, he is on it. Rather than declaring that there is no race, gender, age, or infirmities, he is raceless, genderless, ageless, and infirmity free. He is, perhaps, the very person typing those messages to us. As construed by MCI, then, the Internet offers us the limited opportunity to *pass* as this *fictional* unmarked white male. The text-only corporate disembodied trademark moves us from being marked to being unremarking and unremarkable (on television, at least). That is, if trademarks have traditionally, as Lauren Berlant claims, offered consumers a prosthetic body—a body to take on in public, yet still a body immersed in commodity culture—MCI offers a prosthetic identity that mimics *the original, unrepresentable* prosthetic identity (the fact that MCI must offer an image of a white male, if only to show him ignoring "us," reveals the impossibility of "pure textuality," of pure mastery, as well as the privilege still inherent to this white male placeholding position). Through its play of regularly consumable yet unsatisfactory bodies, through its textual traces and this fleeting white male placeholder, MCI merges the "equality" that supposedly stems from mass consumption with the supposed subject of the bourgeois public sphere, who writes and argues rather than merely consumes. So, if Habermas condemns the market for destroying rational-critical debate by replacing it with consumption (a destruction caused by the very notion of bourgeois private property that also enabled it), MCI offers a way to *buy* oneself back into the realm of rational-critical "debate," which is now redefined as a *marketplace of ideas*.[14] At this moment of unity—of the fixing of formal equality with no apparent cost— "we" agree, "Is this a great time or what?" Technology makes it possible to believe once more in liberal and consumer equality.[15]

14. See "The Blurred Blueprint: Developmental Pathways in the Disintegration of the Bourgeois Public Sphere," in Habermas, *Structural Transformation*.

15. Jodi Dean, citing Slavoj Žižek, argues that the "democratic attitude is always based upon a certain fetishistic split: *I know very well* (that the democratic form is just a form spoiled by stains of 'pathological' imbalance), *but just the same* (I act as if democracy were possible)" (*Publicity's Secret: How Technoculture Capitalizes on Democracy* [Ithaca, NY: Cornell University Press, 2002], 11). This fetishistic split

"Is this a great time or what?" seeks to unify more than the mass and the public subject; it also seeks to unify the nation by fostering historical amnesia and by providing a future alibi. These happy, shining, cutoff faces represent a diverse yet unified United States that opposes the view of the United States as dangerously close to disintegrating into "special interest" groups, that opposes the view that the public sphere has disintegrated into divisive spaces. "Is this a great time or what" as a rhetorical question, and race, age, gender, and infirmities as skin-deep, all erase civil inequalities and civil unrest. According to Berlant, "The trademark . . . [helps] to produce the kinds of historical amnesia necessary for confidence in the American future"; but MCI's trademark goes one step further, enabling future amnesia as well.[16] In the absence of physical evidence to the contrary, it encourages participants to imagine or assume that their audience/fellow surfers are/will be diverse. The mantra "there is no race" serves as an alibi, enabling one to turn a blind eye to demographics for according to this logic, acknowledging race is itself racist. Further, through this ad campaign and campaigns like it, telecommunications companies claim to create utopia/amnesia by privatizing civil rights. There is no need anymore for battles over discrimination because the Internet can guarantee those rights the state has not been able to provide. The "government-free" Internet makes disgruntled faces happy. Thus exploiting the otherworldly dreams that support cyberspace, these corporations offer an unearthly solution to inequality, selling one of the most compromising media to date as freedom.

---

is ideological for, according to Dean and Žižek, ideology operates through one's practices and fantasies (which also suture) rather than one's beliefs. Dean thus maintains that "the concrete materialization of publicity in contemporary techno-culture, a materialization incited by the lure of the secret and the fantasy of its revelation, replaces the fetishistic split with the conviction that democracy is possible; in other words, the knowledge that democracy is not possible is eliminated, replaced by the sense that new technologies enable full access, full inclusion, full exchange of opinions, and so forth" (11–12). Importantly, as the MCI commercial reveals, this belief is limited to technological spaces, and the Internet—based on a most nontransparent technology—does not simply materialize publicity.

16.   Berlant, "National Brands," 188.

Digital Dividends

This is all about self-interest.... There is nothing wrong with self-interest, as long as it is enlightened, long-term self-interest.

—Vernon J. Ellis, international chair for Anderson Consulting and member of the World Economic Forum Task Force studying the Digital Divide

MCI's 1997 "Anthem" represented and still represents a virtual state of affairs.

In 1997, the Internet was not populated by the demographic this commercial and others of its genre portrayed. *Falling through the Net II*, a U.S. government report analyzing telecommunications use in 1997, revealed that White households were more than twice as likely (40.8 percent) to own a computer than Black (19.3 percent) or Hispanic (19.4 percent) ones.[17] According to government figures for August 2000, Asian American and Pacific Islander households had the greatest Internet penetration at 56.8 percent, an increase of 20.8 percent from 1998; White households were second at 46.1 percent, an increase of 16.3 percent from 1998. Even though the rate of increase in Black households exceeded that of Asian American ones (201 percent versus 158 percent) and Hispanic households exceeded that of White ones (187 percent versus 155 percent), Black and Hispanic households were far less likely to have home Internet access (23.5 percent and 23.6 percent respectively) in 2000. According to the 2000 report,

although 56.8 percent of Asian American and Pacific Islander households had Internet access, only 49.4 percent of persons in that group were using the Internet. In contrast, the rates of personal use were higher for Whites and Blacks than their household connection rates. Among Whites, 46.1 percent of

17. National Telecommunications and Information administration (NTIA), *Falling through the Net II*, July 1998, ⟨http://www.ntia.doc.gov/ntiahome/fttn00/falling.htm⟩ (accessed February 26, 2001). During the Clinton administration, the NTIA produced a series of four reports with the subtitles: "A Survey of the 'Have Nots' in Rural and Urban America" (1995), "New Data on the Digital Divide" (1998), "Defining the Digital Divide" (1999), and "Toward Digital Inclusion" (2000).

their households had online connections but 50.3% of Whites were Internet users at some location. The gap was even larger for Blacks: only 23.5% of their homes were online, but 29.3% of Blacks were Internet users. Only for Hispanics were the two percentages essentially the same at 23.6% and 23.7%, respectively.[18]

The growing disparity in real numbers between White and Asian American and Pacific Islander households versus Black and Hispanic ones, despite larger rates of increase in the latter groups, clearly shows the continuing effects of historical inequalities. If all groups continue to increase at these rates, the "unwired" races will never catch up, although we must remember that these figures do not coincide with actual Internet use but rather with the percentage of households that follow the corporate ideal.[19] Indeed, the definition of *digital media* solely in terms of computer use, as Alex Weheliye has argued, erases African American uses of technology and "whitens" cyberstudies.[20]

Countering corporate scenes of empowerment with digital divide statistics, however, is hardly effective, for seemingly contradictory narratives about digital empowerment and disempowerment coexist nicely. Cisco Systems, for instance, ran its ads, in which people of color all around the world happily offer statistics about future Internet usage (claiming they are "ready"), at the same time as it devoted corporate energy toward battling the "digital divide." Viewing this disconnect as discrediting these scenes of empowerment misreads the purpose of this commercial and the commercials of this genre, which dominated the market from 1997–1999. These corporate scenes of empowerment did not seek to get more raced others on the Internet (after all, these others urge viewers to enter *their*

---

18.   NTIA, *Falling Through the Net: Toward Digital Inclusion*, October 2000, ⟨http://www.ntia.doc.gov/ntiahome/fttn00/falling.htm⟩ (accessed February 26, 2001).

19.   For more on this, see Anna Everett, "The Revolution Will Be Digitalized: Afrocentricity and the Digital Public Sphere," *Social Text* 20, no. 2 (Summer 2002): 125–246.

20.   See Alex Weheliye, "Feenin: Posthuman Voices in Contemporary Black Popular Music," *Social Text* 20, no. 2 (Summer 2002): 21–47.

utopia); rather, they sought to convince "the general public" (in particular, business investors—these commercials played nonstop during CNN's *Moneyline*) that the Internet was a safe and happy place. In 1996, the debate over cyberporn saturated public dialogue; films and news features portrayed the Internet as dangerous, and depicted transmitting one's credit card online as an invitation to identity theft. The e-commerce revolution began in 1997. Clearly, not everyone who viewed these commercials believed the Internet a virtually realized utopia, and these advertisements did not single-handedly change public perception. They did, however, help to transform public debate over the Net. After the "Net as Racial Utopia" explosion, the debate centered on the question, To what extent does the Internet allow for democratic exchange and equality? and not, To what extent is the Internet a pornographic badlands or lawless frontier? The former induces commercial transactions far more than the latter.[21]

Corporations also have no problem with the digital divide because they use the disparity between potential and actual empowerment to insinuate themselves as "the solution." By defining technologically produced racial equality as the "ideal," they argue for increased technology adaptation until such racial (consumer) equality is reached, effectively giving themselves an unending "mandate." This mandate to eradicate inequality begs the question, Why exactly is Internet *access* valuable? Indeed, narratives of the digital divide and digital empowerment form a circle that circumvents questions about the value of information, or the value of access alone, since the Internet—redefined through issues of social justice— becomes inherently valuable and desirable.[22]

---

21. The Net utopia explosion did not entirely dispel fear about the so-called pornographic nature of the Internet but rather dislodged it temporarily. These two portrayals serve as the poles between which public perception of the Net swing. Both poles effectively screen the fact that vulnerability and publicity are *constitutive of* the system: one by claiming all vulnerability to be accidental, and the other by displacing this vulnerability onto certain *content*.

22. Faced with the economic slowdown in 2001, members of the U.S. Congress looked to the digital divide for investment opportunities. Rep. Barbara Cubin (R-WY), for instance, introduced a bill in spring 2001 that would provide regulatory relief for all but the biggest telecommunications companies—so they could build more broadband in rural areas.

Government and intergovernmental agencies reiterate this corporate position with few alterations. The 2000 UN secretary general's report on information technology understatedly admits that "even in developed countries, such as the United States of America, a certain time lag was observed between the start of the information revolution and its verifiable impact on the economy, in particular on growth." Although this report dampens the rhetorical extravagances of another UN report by a panel of experts, which urged developing nations to catch the "Internet Express" and get into the "digital age," the secretary general's report cautions against caution, since "the inexorable logic of the emerging knowledge-based global economy and society emerge to make ICT [information and communication technologies] the best hope for developing countries for leveraging their potential and for integrating into the global economy."[23] Submitting to the emerging knowledge-based global economy means enforcing intellectual property laws, selling state-owned telecommunications networks, developing digital sweatshops, and being content with equitable consumer access/production of local *content*. Seeing public service consortia as a "first step," the secretary general argues, "Governments of developing countries and countries in transition ... need to nurture and support the private sector by providing institutional support, reducing barriers to entry, developing sources of financing and helping create and expand markets through tax incentives and export promotion zones etc."[24] Although not explicitly stated in this report, "nurture and support" of private industry means selling national telecommunications to foreign corporations. The United Nations, for instance, lists Estonia as a successful model:

Estonia progressed within a decade from virtually no connectivity in 1991 to now being one of the most connected countries in the world. Today all

---

23.   United Nations, "Development and International Cooperation in the Twenty-first Century: The Role of Information Technology in the Context of a Knowledge-Based Global Economy—Report of the Secretary-General," May 2000, 6, ⟨http://www.un.org/documents/ecosoc/docs/2000/e2000-52.pdf⟩ (accessed February 26, 2001).

24.   Ibid., 24.

schools have been connected to the Internet, 80 percent of bank transfers are made over the Internet. 28 percent of the population is connected to the Internet either at home or at work, and dial-up service is the least expensive in Europe. "Smart cards" are being introduced for use for most services requiring interaction with the public administration, hospitals, public transportation, public telephones, etc. These results were achieved through a concerted national effort that was based on several strategic elements: the newly independent country; belief that ICT could help bridge the gap between poverty and wealth and encourage the rural population to remain *in situ* because it felt connected and a part of the wider world; and a depoliticization of the connectivity issue by entrusting a specially created NGO ("Tiger Leap Foundation") with determining which communities would be allotted government monies for the purchase of hardware and software. An important element of this approach was the stipulation that recipients were required to pay 50 percent of the cost, thereby creating the sense of ownership.[25]

Glaring omissions in this list of "causes" for Estonia's success—and indeed, missing throughout this report—are the facts that Estonia created "the requisite infrastructure through a concession agreement with Swedish and Finnish telecommunications operators by which they modernized the telephone network in exchange for profits from the telecommunications business," and that outsourcing has been key to driving down the cost of computers.[26] Moreover, the list of so-called improvements brought about by Estonia's telecommunications modernization begs the question, Exactly why are smart cards and Internet bank transfers indicators of progress? (unless, of course, the goal is better surveillance and easier corporate transfers).

To be clear, privatization is not always bad, nor is government control always good: state-owned telecommunications networks are not automatically more public than privately owned ones (indeed, when the Internet was under the control of the U.S. government, it was mainly used for

---

25. Ibid., 6–7.

26. United Nations, "Report of the Meeting of the High-Level Panel of Experts on Information and Communication Technology," May 2000, 14 ⟨http://www .un.org/documents/ecosoc/docs/2000/e2000-55.pdf⟩ (accessed February 26, 2001).

military and academic purposes). More important, "information" and "knowledge" (I put these terms in scare quotes because these key words are assumed rather than defined) call into question private ownership. As the secretary general's report points out, information and knowledge "cannot be depleted. Their use by one does not prevent their use or consumption by another. They cannot be owned, though their delivery mechanisms can. Selling them entails sharing, not exclusive transfer. Indeed, information and knowledge represent a global public good."[27] In this sense, information (if by this we mean electronic data) is the anticommodity: it cannot be transferred or owned exclusively—if, of course, there is any*thing* to "own" in the first place; because digital media automatically copies what it downloads, the user is by default always "accumulating."[28] Such a technologically based explanation, however, ignores that ghostly immaterial presence that transforms goods into commodities; the perceived exchange *value* of "intel" transforms electronic data into commodities. Electronic data—and at one point dot-coms—reverse Marx's comment that "a thing can be a use-value, without being a value."[29]

Informationology—information is knowledge—is buttressed by intellectual property laws, conflates data with power, and endows values to useless nonobjects. Informationology, an admittedly ugly neologism, refers to the perversion in the will to knowledge specific to computation: the almost religious belief in the value of information, which manifests itself in everything from game shows that reward trivial knowledge to the deluge of "biographies" on cable television, from cyberpunk hero/ines who make money by selling interesting tidbits of information to obsessive-compulsive sysops who archive everything and then occasionally sell these archives. Informationology depends on a logic of scarcity that belies the

27. United Nations, "Developmental and International Cooperation," 9.

28. This constant accumulation leads to a situation in which people tend to store data, which is often useless, on the off chance that it might later be useful. Norbert Wiener, in *The Human Use of Humans: Cybernetics and Society*, (Cambridge: DaCapo Press, 1954) argues that information cannot be commodified because it is nonconservative (116).

29. Karl Marx, *Capital*, trans. Ben Fowkes, vol. 1 (New York: Penguin Books with New Left Review, 1976), 131.

United Nations's celebration of knowledge—information becomes valuable when it is portrayed as belonging or restricted to certain persons; information becomes valuable when language itself becomes "owned." This value depends mainly on intellectual property laws, which turn ideas and language sequences into property, and assume the universal applicability of knowledge, and partly on technology that "fixes" free copying. The success of intellectual property laws shows how flexible capitalism can be in what Mark Poster has called capitalism's "linguistic turn."[30] Thus, the assumption that information technology represents a global public good relies on a naive reading of the current technology that deliberately ignores the intellectual property laws endorsed in this very document.

To be clear, access and local content production are important. Alone, however, they are not enough to redress inequality but enough to sustain it. As the United Nations argues, producing local content does boost local access: the production of a Chinese-language computer interface was key to the explosive growth in Chinese computer and Internet usage. Yet ending with indigenously produced Web pages sustains English-based programming languages and operating systems as *universal* knowledge—indeed, local access to indigenous Web pages supports ICT as universally valuable. As well, focusing on local access overlooks questions of infrastructure and connectivity: bandwidth does make a difference, and the North's insistence that nations in the South can "leapfrog" them by employing wireless networks screens questions of security and stability. Fiber-optic networks are far more secure than wireless stations and satellites (as mentioned previously, the United States and the United Kingdom bombed Iraq in 2001, as it was completing a Chinese-engineered fiber-optic network).

The critical questions are: What would the United Nations's list of successful nations look like if success entailed the development of skills necessary to affect the Internet's infrastructure or to design software/hardware? How would the definition of successful change if one analyzed the impact of ICT on class structure *within* countries, rather than assume everyone benefits equally from it? Economically, ICT can help; it can also

---

30.   For more, see Mark Poster, *What's the Matter with the Internet?* (Minneapolis: University of Minnesota Press, 2001).

be key—and has been key in all "kinds" of nations, from the United States to Serbia—to disseminating independent news coverage (as well as disseminating propaganda and misinformation). Understanding the ways in which global telecommunications networks can be used to foster social and economic justice, though, entails realizing that ICT does not automatically mean more democracy.[31] To discuss "bridging" the digital divide solely in terms of Internet access—as though serious disparities in types of access did not exist, as though access was enough to dissolve inequity—is disingenuous. These efforts would create "junior users" not unlike "colonized" subjects who were structurally dependent on knowledge from the "motherland." The dominance of the English-language-based programming languages concentrates programming jobs in English-speaking nations (hence the phenomenal growth of the software industry in parts of India). The dominance of English, combined with the overwhelming U.S. predominance in portal sites and the dominance of northern telecommunications companies, all combined with the concentration of capital within the North, makes these enlightened bridging solutions attempts to solidify, rather than reduce, electronic disparities.

These solutions also solidify nonelectronic disparities by obfuscating and exacerbating the problem of debt refinancing. The G8 nations, at their 2000 Okinawa meeting, adopted an informational technology charter intended to bridge the technology gap between rich and poor nations. The prime minister of Japan, Yoshiro Mori, announced that Japan would also commit $12 billion in loans and $3 billion in grants over five years to information technology initiatives in the developing world. The then corporate chair of the then respectable Anderson Consulting admitted that encouraging ICT use in developing and in transition nations was self-serving, but excused it as "enlightened" and "long-term" self-interest. Prior to the G8 summit, however, the African Diplomatic Corps in Tokyo, the Ministry of Foreign Affairs of Japan, and the United Nations University organized a "Global Partnership for Peace, Progress, and Pros-

---

31. The fact that open source is viewed by China as a way to get around the Microsoft monopoly on operating systems belies U.S. libertarian conflations of open source and freedom.

perity: A Message from Africa" to give the G8 countries a sense of African needs. In their conference statement, which is not an unmediated message from "Africa," they emphasized poverty and debt relief. The G8 summit largely ignored this message, even though it could have intertwined information technology investment and debt relief. Indeed, the United Nations made a miniscule gesture toward uniting these two by suggesting 1 percent of a nation's debt be forgiven, if these funds go toward informational infrastructure.

Within the United States, solutions to the digital divide similarly concentrate on access to the Internet, rather than the tools and the skills needed to transform it and similarly erase class difference. For instance, Cheskin Research, in its 2000 *The Digital World of the US Hispanic*, portrays the digital divide as a consumer issue:

With the emergence of the Internet as a vehicle for the new economy, Hispanic consumers represent an untapped market in this new digitally connected world. While today's interest in the Hispanic market is not unexpected, the lack of past interest on the part of technology producing companies is. Software and hardware manufacturers have paid little attention to this potentially lucrative segment. The result is a disparity in household computer ownership and access to the Internet between general and Hispanic markets. This disparity has translated into what has come to be known as the Digital Divide.[32]

According to Cheskin, the digital divide—understood as the disparity in computer ownership between various U.S. ethnic/racial groups—stems from software and hardware manufacturers' lack of marketing, for the "consumption of technology is driven by information, and the Hispanic consumer has been routinely bypassed by traditional marketers."[33] Most bluntly, corporations have failed to make "the US Hispanic" realize that s/he needs technology. The bulk of this report profiles the "attractiveness" of *the* Hispanic consumer: computer ownership has increased by 68 percent since 1997 within Hispanic households (as opposed to 43 percent

---

32. Cheskin Research, *The Digital World of the US Hispanic*, April 2000, 1.

33. Ibid.

among the general U.S. population, according to a 1999 National Public Radio/Kaiser/Kennedy School survey); 75 percent of Hispanic households that own computers also own credit cards, earn a median income of $40,000, and tend to be households of single males (as opposed to a median income $30,000 for mainly female-led households that do not own computers). Although Cheskin's research certainly exposes disparities between racial groups within the same economic bracket and the importance of desire to information consumption, it elides the relationship between race and class. Of all the people Cheskin interviewed, 46 percent said that they did not own a computer because it is too expensive. This 46 percent of "the Hispanic" will not be addressed by strategies to make the Web more attractive to middle-income people of color—tactics taken up by many e-commerce Web sites, such as http://www.ebony.com.

Through these consumer-based tactics, the Internet *proliferates* race. This proliferation of race as a consumer category also constructs race or ethnicity as a category to be consumed: it encourages one to celebrate, or to identify with, another race by indulging in the same "authentic" pleasures. According to Jennifer Gonzalez, this form of consumption as passing, where one takes on a marked body rather than an unmarked one, stereotypes and fetishizes. In her reading of "virtual worlds," Gonzalez argues that the fantasy of "taking on another body" merges together the postmodern subject with the transcendent subject of old to create a new cosmopolitanism.[34] Through this new cosmopolitanism, one avoids the complex subjectivity of the other: the postmodern (virtual) subject appends various racial features to itself in order to "pass" as the other, with no regard to historical specificity or social process. Significantly, this phenomenon is not limited to cyberspace; the most banal and prevalent example of this is multiculturalism as a form of "taste-testing" (for instance, "Honey, let's eat Chinese tonight"). The most extraordinary example of this is A&E's portrayal of Jeffrey Dahmer as a "multicultural" mass murderer because he killed and feasted on people of color (after he decided he was a cannibal). This idea of consuming what the other con-

34. See Jennifer Gonzalez's "The Appended Subject: Race and Identity as Digital Assemblage," in *Race in Cyberspace*, eds. Beth Kolko et al. (New York: Routledge, 2000), 27–50.

sumes, or literally consuming the other—desiring what the other desires, desiring to be what the other desires—leads to the increasing presence of "racial" categories as pornographic ones. The Internet has not stopped the display of raced bodies—on the contrary, race has become entrenched as a pornographic database category (one of Marty Rimm's new categories). As mentioned in chapter 2, googling "Asian + woman" produces more pornographic sites in its top-ten hits than one on "pornography." But race is not simply a pornographic category. As Anna Everett notes communities of color began using the Internet years before the Internet as racial utopia explosion and for purposes other than declaring there is no race.[35]

Regardless, many English-language "Asian" porn sites make clear this slippage between race as a consumer category and fetishistic passing, since they construct "authentic" ethnic subjects through "fetish" or "exotic" desires, interpellating their users as "samurai" or "papa san."[36] For instance, the introduction to asiannudes.com in 1999 encouraged surfers to become samurai:

You are welcome to our dojo! Look no further, traveler. You have found the Clan of Asian Nudes, filled with gorgeous Asian women in complete submission. Take them by becoming a samurai. Our dojo houses the most incredible supermodels from Japan, Vietnam, China, Laos, and San Francisco's Chinatown! Their authentic, divine beauty will have you entranced nightly. New girls are added almost every day, their gifts blossoming before you on the screen.[37]

Whether the viewer is Asian, Asian American, or non-Asian, the site seeks to make one feel like all-powerful samurai, an all-powerful user—spectator rather than spectacle. This open invitation to "become samurai" reveals the mediated nature of identification since Asians and Asian

---

35.   See Anna Everett, "The Revolution," 125–246.

36.   The introduction to xxxasians.com reads: "The Streets of Beijing to the Red Light District of Tokyo. We've Picked out for You Only the Best Girls Doing the Hottest Nastiest Sex Acts. They Are All Waiting for "YOU" Papasan What More Could You Ask For?" (accessed February 9, 2001).

37.   *Asian Nudes*, ⟨http://www.asiannudes.com/tour1.html⟩ (accessed April 1, 1999).

Americans too must pass as samurai (given that most of Hapa porn star Asia Carrera's fans are Asian/Asian American, a significant portion of these sites' visitors is also probably Asian/Asian American. This complicates any simplistic reading of the "Orientalism" involved on these sites and the significance of male white figures on them). Clearly, "passing as a samurai" differs from passing as an invisible white male, but both require concurrent identification and misrecognition as well as the objectification of others. If pornography in general, as Linda Williams has remarked, has been linked to the "frenzy of the visible"—the increasing desire to "see" and "know"—these Web sites reveal the link between pornography and the frenzied display of authentic ethnic knowledge, authentic ethnic information.[38] For instance, xxxasians.com claims it has "Asian Sex Shows! Better Than Amsterdam. We hit the sides streets of Singapore and the remote hidden away places in Bangkok to find the best for you.... Asian Porn at its best!! 100% legit Asian films. Don't be fooled by these other Asian/American sites. We have the best LEGIT Asian porn, shot straight from the backwoods of Asia." The privileged position of the viewer depends on the reduction of authentic others to flesh, to flesh made information. The fantasy here, as in most mainstream pornography, is the fantasy of catching the authentic other unawares (in the backwoods of Asia), so that it reveals its secrets to the viewing subject. These Asian porn sites, which like all Web sites endlessly circulate the *same* pictures, offer the lure of newness or breaking news. "Want something new?" asiannudes.com asks, "Having the largest Asian data base, we add new girls to our site every day! Other sites add pics weekly or monthly. But not here! At Asian Nudes, we present you with new girls every day, GUARANTEED!"

As argued in the previous chapter, although pornographic Web sites' content may be directed at making their viewers seem all-powerful, pornography's elucidation of visceral responses—the ways it moves its viewers—hardly enables "supercontrol." As well, porn sites are usually anything but empowering. They rewrite browser defaults through javascripts so that new windows open when you try to close them, so that your back button takes you to another porn site. They deluge your inbox; they collect IP information; they track their visitors. On porno-

---

38. Linda Williams, *Hardcore*, 34–57.

graphic sites (and commercial sites more generally), users experience the greatest disparity between their perceived and actual level of control. In order to compensate for the ways in which interactivity breaches the "self," in which the electronic self emerges through the call of another (a call that unlike linguistic calls, cannot be "read" or "heard"), dreams of superagency emerge.[39]

Refusing Markets

Race may be a pornographic (consumer) database category, but this does not mean that all references to race on the Internet are racist or pornographic, that all pornography simply reinforces racism, or that consumption is always racist. As Ernesto Laclau and Chantal Mouffe have asserted, "Interpellated as equals in their capacity as consumers, ever more numerous groups are impelled to reject the real inequalities which continue to exist. This 'democratic consumer culture' has undoubtedly stimulated the emergence of new struggles which have played an important part in the rejection of old forms of subordination, as was the case in the United States with the struggle of the black movement for civil rights."[40] Consumer equality—and the demand for autonomy both spurred on and denied by its promise—is part of the "democratic adventure"; it is an extension of the logic of equivalence beyond what has traditionally

---

39.   Mark Poster, in his provocative analysis of "virtual ethnicity," argues that ethnicity is the product of many everyday practices and is itself constructed. Thus, comparisons between virtual ethnicity and "real ethnicity," which portray real ethnicity as somehow fixed and lost, create ethnicity as fixed and lost. This does not mean that the two terms are equivalent, however; as opposed to "real life," the Internet enables an "underdetermined" subject to emerge—a subject who is implicated within the circuit of the Internet, who is not abstracted from technology because the technology itself is not an "object" but rather a social place. This means that virtual ethnicity enables a far more fluid subject to emerge. Although this analysis is important, it ignores the ways in which the user's fluidity is compensated for by dreams of superagency. Moreover, it is not simply that analyses of virtual ethnicity perpetuate real ethnicity as solid but that representations of virtual ethnicity and the call to pass do so as well.

40.   Ernesto Laclau and Chantal Mouffe, *Hegemony and Socialist Strategy: Towards a Radical Democratic Politics*, 2nd ed. (London: Verso, 2001), 164.

been considered the political and public. If "the public/private distinction constituted the separation between a space in which differences were erased through the universal equivalence of citizens, and a plurality of private spaces in which the full force of those differences were maintained," the demand for new social rights seeks to dissolve this distinction and explode the political by turning subordinate relationships into antagonistic ones.[41] The Internet as racial utopia rhetoric seeks to eradicate antagonism by offering a space of virtual equality and autonomy, and by reworking the antagonism so that domination stems from one's very body. In doing so, it makes one's body something to be consumed—it makes one's race a commodity in order to erase it. This corporate hijacking of democratic logic works to ensure inequality. Thus, effective antiracist uses of the Internet must not commodify or erase race.

The software art projects by the U.K.- and Jamaica-based "digital collective" Mongrel exemplify such a strategy. Mongrel's search engine Natural Selection, for instance, ties antiracist Web sites to racist searches. As Graham Harwood, part of the Mongrel core, explains to Matthew Fuller, an artist/programmer/activist/mongrel who worked with Harwood on Natural Selection:

Well basically, it's the same as any other search engine. The user types in a series of characters that they wish to have searched for. The engine goes off and does this and then returns the results. If you're looking for sites on monocycles, that's what you get. If you're looking for sites on elephants, that's what you get. As soon as you start typing in words like "nigger" or "paki" or "white" you start getting dropped into a network of content that we have produced in collaboration with a vast network of demented maniacs strung out at the end of telephone wires all over the place. The idea is to pull the rug from underneath racist material on the net, and also to start eroding the perceived neutrality of information science type systems. If people can start to imagine that a good proportion of the net is faked then we might start getting somewhere.[42]

41. Ibid., 181.

42. Quoted in Matthew Fuller, "The Mouths of the Thames," ⟨http://www.tate.org.uk/webart/mongrel/home/faqs/ns.htm⟩ (accessed February 26, 2001).

Natural Selection is an elegant hack on search engines: it uses another database to run its nonracialized queries, illustrating nicely Mongrel's tactic of "taking on the media by mounting it from the rear" (as does their commissioned reworking of the Tate Museum site, which oscillated between loading in front of and behind the official one). Designed to dispel the specter of "racist material on the web" through parody rather than censorship, Natural Selection highlights the duplicity of electronic communications: advertising Natural Selection as a way to "stop you smearing skin lightener on your computer," Mongrel and its associated network of "demented maniac" content providers offer parodies of racist Web sites, such as Goldhorn's Racially Motivated Fuck Fantasia, as well as sites, such as By Bad Boy Byju's Aryan Nations, that reveal how "pure English" was always already overrun with the language of the colonized.[43] These sites attack the reliability and the authenticity of online representations of raced others. If the user as superagent emerged partly as a means to blind users to the constitutive duplicity and unreliability of "information," this attack belies the value of the information superhighway and offers an opportunity to rethink the Internet as means for publicity.

The Natural Selection Web sites are not simply antiracist. As Fuller remarks, "If people are going to check it out, they need to be looking for more than a punchline, or a nice neat 'anti-racist' or 'multicultural' solution."[44] They should also be prepared, in some cases, to be confronted with pornographic images that expose the thin line between white supremacist fetishism and gay pornography, to listen to white supremacist punk spliced together with black nationalist rap, and to be interrogated by javascript alerts: to make race "hard to consume," the Natural Selection Web sites deny the "distance" needed for a color-blind subject to emerge. They also refuse to offer authentic images of others as a way to counter racist stereotypes that are perpetuated online—they refuse to

43.   As Matthew Fuller explains: "Along with porn, one of the twin spectres of 'evil' on the internet is access to neo-nazi and racist material on the web. Successive governments have tried censorship and failed. This is another approach—ridicule" (⟨http://www.mongrelx.org/Project/projects.html⟩ [accessed February 26, 2001]).

44.   Fuller, "Mouths of the Thames."

confess their (sexual/ethnic) "truth" to the user for its edification. As Rey Chow, drawing from the work of Foucault, has argued in "Gender and Representation," self-representations do not get us out of the bind of representation since they too operate as "voluntary, intimate confessions" that can buttress power.[45] Rather than presenting themselves as authentic "amateurs" outside representation, these sites interrupt the *pleasure* of knowledge, the pleasure and the mastery of the user.

Dimela Yekwai's "antiviral" site, for instance, juxtaposes the words of "Venus-Fly-Killer" next to the racist words of "Bombarded-Images."[46] Using pronouns such as I and you, these poems establish a personal relationship between these voices and the user as well as between each other, setting up what Yekwai calls a "triple-consciousnessed Afrikan virus." The "Venus-Fly-Killer" section begins with: "Let me introduce myself / I am / $\gg$ Venus Fly Killer $\ll$," whereas the "Bombarded-Images" poem begins with: "Let me introduce myself / you / $\gg$ Black-bastard $\ll$."[47] In the "Venus-Fly-Killer" poem, the narrator's lyrics kill racism, which she figures as flies; in the "Bombarded-Images" section, the white male narrator bombards the user with racist epithets, and explains how the "blessed-race" used genetics to extinguish blacks and create a "lily-white" environment, in which the remaining people live with bubbles on their heads.

In addition to these conflicting addresses, Yekwai's javascript alerts interrupt the user, highlighting "responsibility" (the call of the other precedes the "user" or self). The first alert informs the user and the narrator of "Bombarded-Images," "You can't get rid of me." Once the user clicks "OK," Yekwai's spinning head emerges in a smaller window to the upper-right-hand side of the monitor. The virus that she protects against (her spinning head "accompanies" us through our travels) is racist information: "To / help you in your quest for / truth, For life itself, as I / have stated earlier, I will now / traverse the INTERNET with / you, The highway

45.   Rey Chow, "Gender and Representation," in *Feminist Consequences*, eds. Elisabeth Bronfen and Misha Kavka (New York: Columbia University Press, 2000), 46.

46.   ⟨http://www.mongrelx.org/Project/Natural/Venus/index.html⟩   (accessed February 1, 2001).

47.   Ibid.

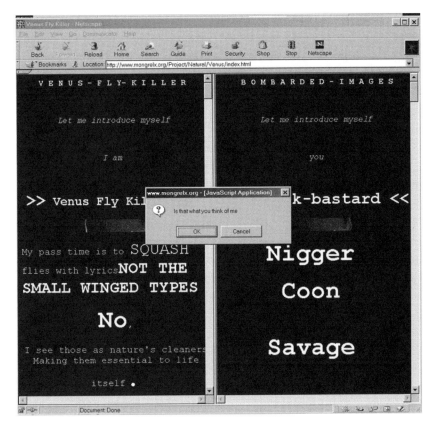

| Figure 3.8 |
Yekwai's "Venus-Fly-Killer" and "Bombarded-Images,"
⟨http://www.mongrelx.org/Project/Natural/Venus/index.html⟩

on which most / of these lies lurk, I can only / alert you to their hiding / places / Make the choice NOW! / YOU HAVE THE / POWER!!".[48] Once her head has emerged, more alerts bombard would-be readers/users. First, they are offered an epithet, such as "Lazy Bums"; in order to continue, they must click "OK." Then, they must answer the question, "Is this what you think of me?" Using the default settings of javascript,

---

48.   Ibid.

this alert offers two answers: "OK" or "cancel" (figure 3.8). If you answer "OK," you are taken to the specific section in the "Venus-Fly-Killer" poem that addresses this racist epithet. If you answer "cancel," you are taken to the section within the "Bombarded-Images" poem that perpetuates the racist term. These alerts leave no position for denial, no way to say no, no way of putting her site into the background. So, if you have the site in one window, but are working in another, your operating system interrupts you and lets you know that Netscape Communicator needs your immediate attention. Yekwai's site is interactive, but in a manner that belies interactivity as user controlled and thus returns interactivity to its earlier meaning.[49]

Another Mongrel project, National Heritage, confronts racism as a global national heritage. Described as an "abortion" on "cyber-civilization," Natural Heritage wages "info-war against the racially-exclusive, US west coast eutopian nonsense" and seeks to take the "future" away from those "who left us out of the past."[50] By insisting on this "heritage," it historicizes and disseminates images of raced others, while at the same time refusing to offer users ethnographic images. National Heritage uses seemingly stereotypical or representative "amateur" racial specimens; however, these images (drawn from Mongrel's Colour Separation project) are in fact composites of numerous "friends" of Mongrel. These images are also offered in a grid that shows them with varying "colors." In the National Heritage installation/software, the user must spit on these images in order for them to "tell" their stories (spitting on these images also changes and produces their mask). This spitting not only establishes these persons' experiences as bruised—and relationships in general as conflictual rather than peaceful—it also implicates the user as part of a racist society. Yet through this spitting, understanding and mask changing can also emerge. Through this denial of indexicality, Mongrel seeks to bring out how

---

49.   Interactivity stemmed from artificial intelligence, from the realization of the limits of human programming capabilities. It involved giving over to the machine tasks humans could not perform. For more on this see, Wendy Hui Kyong Chun, "On Software, or the Persistence of Visual Knowledge," *grey room* 18 (winter 2005): 26–51.

50.   ⟨http://www.mongrelx.org/Project/projects.html⟩.

constructions of race in the form of mental images are much more than simple indexes of biological or cultural sameness. They are the constructs of the social imagination, mapped onto geographical regions and technological sites.

These fabrications of race have traceable links to historically specific relations, from those informing the experience of slavery, migrant labour, colonisation, to those affecting friendship and family life. Racial images are pregnant with the social and political processes from which they emerge and to which, in turn, they contribute, and images of different races articulate the political and economic relations of races in society.[51]

Through fabrications of "people that never existed," Mongrel insists on the importance of historical and economic contexts. By exposing the duplicity central to digital imaging, the collective exposes the duplicity central to racial stereotyping. So, if Jennifer Gonzalez argues that the Undina project preserves racism by starting from these so-called stereotypical images, Mongrel attacks the premise of these images, showing that such constructions can only be fabrications.[52]

The mask portrays interactions as always mediated. According to Mervin Jarmin, another core member of Mongrel:

I believe the mask to be one of the most defining aspect of the whole project in more ways than one; the mask represent the mask that I always have to wear at the point of entry into Britain, it represent the mask that I wear repeatedly as I go about my everyday activities in this lovely multicultural state.... And then it also represent the mask that mongrel has to wear in sourcing resources for the project. So you see the whole National Heritage project is a constitution of the mask.[53]

National Heritage reveals the mask—the state of passing—rather than stereotypical raced selves as the default. Just as the figure of the mask

---

51.   Ibid.

52.   Jennifer Gonzalez, "The Appended Subject: Race and Identity as Digital Assemblage," in *Race in Cyberspace*, eds. Beth Kolko et al., 27–50.

53.   Quoted in Fuller, "Mouths of the Thames."

———

reverses the relation between stereotype and passing, Mongrel's interfaces and software reverse the usual system of software design: it produces interfaces and content that are provocative—even offensive—in order to reveal the limits of choice, to reveal the fallacy of the all-powerful, race-free user. Mongrel also produces software tools such as Linker, which are extremely easy to use, and works with historically "unwired" communities to produce beautiful digital projects.

Mongrel's projects also play with the relationship between software and ideology in order to make us question the reduction of race to a database category. The collective's HeritageGold software highlights the relationship between software and ideology, software and race beautifully. Through rewriting the standard menus of Photoshop 1.0, Mongrel addresses the politics of changing color, of passing. For instance, under the "social status" ("image" in the original Photoshop menu) option, it translates the RGB color setting to Middle Class and the Index setting to aspiring. So, in order to apply a social filter to the image (assimilate, add more cash), one must first make the image Middle Class (see figure 3.9). The image channels are AAA, Aryan, Asian, and Afro. To save, one "births"; to close, one "kills." Page setup is "immigration setup"; printing is "migrating"; one opens families instead of files; one copies and pastes skin, and fills in flesh wounds; one defines breeds and patrimonies (if the selected area is too large, one cannot define a breed because the area is too big to be a ghetto). Mongrel's "historical relations" option (which allows one to apply various masks to the images) is particularly insightful. The slavery function (which transforms the image into black and white, and lightens it) adds black.female.lut using the Aryan channel. This simple hack of Photoshop thus insightfully and provocatively manipulates the resonances between race and software in order to make clear the costs and assumptions behind the rhetoric of the Internet as race free. Making explicit the parallel between race and software enables a response to the simultaneous erasure and commodification of race and software.

Mongrel's projects deliberately and insightfully attack prevalent notions of "interface" driving myths of computer access as equality. Mongrel's project of spreading critical literacy about the Web through projects that move along the same trajectory as racist terms, brilliant as it is, also runs into the problem associated with all parodies—namely, the question of audience. There is no guarantee that readers will recognize

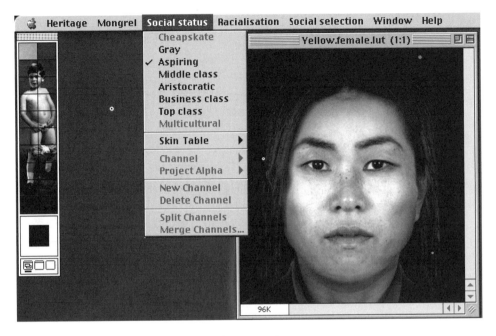

| Figure 3.9 |
Mongrel's HeritageGold.

the Mongrel sites as parody rather than the "real thing." This difficulty is exacerbated by the fact that people surf rather than read on the Net. How many users, for instance, will read Yekwai's entire poem? How many of them will recognize Critical Art Ensemble's bio.com's site as a parody (arguably, the "real" eugenics sites to which it links are the most terrifying). Making Natural Selection accessible to its so-called target audience—those genuinely searching for racist terms—is complicated by the fact that Natural Selection is currently off-line (search engines view Natural Selection as a hack and refuse it access). Regardless, search engines do index these sites, and perhaps this indexing is the best way to ensure success. If a search on "Asian + woman" on Google could bring up Goldhorn's Racially Motivated Fuck Fantasia, then perhaps we would be getting somewhere. Kristina Sheryl Wong's site Big Bad Chinese Mama, which satirizes mail-order bride and Asian porn sites, takes such an approach. As she explains to the *Village Voice*, Wong began copying porn site's metatags in order to rank higher on search engines: "They [metatags] were huge, and would be jammed full of search terms like 'blow job,' '69,' 'ass,' and 'dutch'—I don't even know what 'dutch' is.... Now I love to check my statistics and see what people typed in to find my site. One time, it was 'Eskimo-fucking Cambodian women.'"[54]

All these sites, however, raise the question, What next? Harwood, discussing Colour Separation, asserts that "in this work as in the rest of society we perceive the demonic phantoms of other races. But these characters never existed just like the nigger bogeyman never existed. But sometimes ... reluctantly we have to depict the invisible in order to make it disappear."[55] The question that remains is, How exactly will these depictions make the invisible disappear? Clearly, the grid structure of Colour Separation, which shows the various transformations and their masks, troubles "natural" assumptions about race and makes the viewer pause, and perhaps even pause in the ways that Mongrel intends. But given the increasing tendency to view culture, rather than biology, as the term that creates irreconcilable differences—a type of racism Etienne Balibar calls "neoracism"—the insistence on biological fluidity is not enough in and of itself.[56] Taken as a whole, Mongrel's projects interrogate both the biological and the cultural; the question is, How can one highlight the whole given that surfers move from part to part?

Further complicating this work is the question of reincorporation. The presence of Mongrel's Uncomfortable Proximity site behind the Tate Museum site does mount the media from the rear, but it also becomes enfolded within the official site itself, revealing the incredible flexibility of what Mongrel calls the "bourgeois sensibility." Uncomfortable Proximity can translate into uncomfortable incorporation, and computer networks exacerbate this tendency since they do not allow for "outsides"—everything becomes yet another window within the same

54.   Quoted in Logan Hill and Thuy Linh Nguyen, "Asian Artists Make Porn Sites Work for Them: Nude Japanese Schoolgirls! Lotus Blossoms! Radical Feminists?" *Village Voice*, August 22–28, 2001, ⟨http://www.villagevoice.com/issues/ 0134/hill.php⟩ (accessed August 25, 2001).

55.   ⟨http://www.mongrelx.org/Project/projects.html⟩.

56.   Etienne Balibar and Immanuel Wallerstein, *Race, Nation, Class: Ambiguous Identities*, trans. Chris Turner (London: Routledge, Chapman and Hall, 1991).

screen. Mongrel's work is within the circuit, no matter how hard it may protest to be outside it. This is not a condemnation, for there are different ways of being "inside," but this fact does pose another concern about National Heritage: making users spit may expose our relation to another's pain, but it also flattens differences between users. Also, making the "faces" speak after being spit on exposes the ways in which the other speaks its truth in response to the demands of the would-be user/subject, but it forecloses the possibility of silence and refusal. It would be intriguing to have one face (besides that of the white male) that refused to speak, no matter how much one spit on it. Such an intraface would bring out more clearly the violence associated with making one speak and also explore the possibilities of silence.

Regardless, Mongrel's projects highlight the fact that the pornographizing of difference does not close the possibility of the Internet as public, as a new and more open means for textual communication with others. It offers us a point from which to begin an analysis of the Internet as a rigorously public medium. To begin this analysis, though, we must explore the consequences and possibilities behind intrusion and disembodiment (albeit disembodiment in a nontextual sense). We must give up fictions of total security sustained by the frenzied display of others, and understand how the public operates through disembodiment and intrusion—and how it is from this disembodiment and intrusion that we emerge as users.

Disembodiment—and not disembodiment as empowerment—has always been part of representative democracy. As Claude Lefort observes, "It is at the very moment when popular sovereignty is assumed to manifest itself, when the people is assumed to actualize itself by expressing its will, that social interdependence breaks down and that the citizen is abstracted from all the networks in which his social life develops and becomes a mere statistic."[57] Thus, democracy, rather than creating individual speakers fully in control of their actions and respected as individuals, reduces citizens to abstractions, whose reactions are measured statistically. This same abstraction takes place in political discussions on the Internet. At the very moment when individuals are assumed to be engaging in public discussion,

---

57. Claude Lefort, *Democracy and Political Theory*, trans. David Macey (Minneapolis: University of Minnesota Press, 1988), 18–19.

their opinions are abstracted and their texts compromised. During the 1998–1999 debate over President Bill Clinton's impeachment, for instance, electronic communications enabled more contact between citizens and their representatives: sites such as Moveon.org and electronically forwarded e-mail petitions made contacting one's representatives as easy as clicking a mouse button. This arguably revealed that the Internet could lead to more meaningful participation by the citizenry—rather than simply registering a vote during elections, citizens were taking a more active and textual role in democracy (it was not simply a vote, it was an e-mail). On the other hand, the sheer number of e-mails sent guaranteed that they would remain largely unread and often crashed representatives' servers. Read or not, however, these e-mails served as a harassing message, whose import was measured by *number*, just as votes. Furthermore, these e-mails inverted the relationship between header and body: given that many of these e-mails were forwarded and thus identical, the subject header came to represent the message. Through the subject headers, these e-mails were quickly divided into pro- or anti-impeachment camps. Lastly, e-mail eradicated the semblance of personal dialogue between representative and represented. Although the signed letter of acknowledgment tried to sustain the fiction of personal contact and concern, the automated e-mail "thank you" exposed the mechanical nature of the entire interchange. Hence, as a public space, or a space for public discourse, cyberspace does not ensure that individuals will be able to fully explain, sell, and control their opinions. It does, however, offer a means by which their words— which are also citations of another's—are compromised, acknowledged, ignored, and assessed.

Moreover, unread electronic texts still function. As mentioned earlier, Electronic Disturbance Theatre's software, Floodnet, allows users to tie up servers (it takes advantage of the fact that most servers cannot handle many simultaneous requests). The troupe first unleashed its software during a "virtual sit-in" on the Mexican and U.S. governments' Web sites in response to the military suppression of the Zapatistas. After a similar attack was unveiled on CNN.com and Yahoo.com, such acts no longer qualified as civil disobedience, but are criminal offenses (a consequence of the Internet going public by being sold to private corporations). Still, such acts reveal how Internet communication can work to belie the marketplace of ideas or commodities.

Most basically in terms of disembodiment, TCP/IP precedes individual users, openly copying and transporting seemingly private requests and messages, and reducing identity (that by which one can be identified) to numerical representations of voltage differences. Again, the notion of Internet communications as private is a software effect, and privacy emerges, if it does, from this public interaction. This privacy effect is expensive to maintain, and requires technical and nontechnical intervention: encryption, regulations against placing indiscriminate packet sniffers, public pressure against companies like Google scanning e-mail messages to include targeted ads, libraries that actively delete user files. Remarkably, even when seemingly private interactions are exposed as open—companies and universities openly declare that they, and not their users, own their e-mail messages—many operate as though their electronic communications were personal. But what would happen if users treated their jacked-in machines and messages as public? In all probability, more users would use encryption in order to protect certain messages (again, privacy follows from publicity), and this increased use would help displace the overriding ideology that secrecy equals evil. User profiling would also be more difficult if personal computers were no longer personal. Users might also be inspired to push the limits of networking, to participate in networks that are not simply server-client.

Systems like Napster, Gnutella, and Free.net that allow users to access each others' files start us on this process. They make explicit and usable what many systems hide—namely, the fact that your networked computer engages in a constant give and take. They do not celebrate the fact that every listener can be a speaker but that structurally every computer, in order to communicate, sends and receives information. To return to Richard Dienst's interpretation of Martin Heidegger cited in the introduction, in order to receive images—or any information—one must send information, one must become also a source of information. Free.net takes this idea of publicity and uncertain public spaces even further since there is no central site. Closing the central Napster Web site effectively shut down Napster; Free.net and other systems that are truly distributed cannot be shut down in this manner. So, if Lefort writes that "power becomes and remains democratic when it proves to belong to no one," the same perhaps can be said for networks, although the "spyware" attached to peer-to-peer software such as Kazaa—spyware that reports your activities to a central site—calls into question any easy conflation of hardware with theory or politics.[58] Again, spyware is only possible because operating systems shield us from networking activity in order to make computers seem more user-friendly, more private. It is only by simultaneously screening such activity and provoking fear over what exactly one's machine is doing that Microsoft can claim that it's better with the butterfly or Earthlink's halo can be construed as comforting. Freedom reduces to security.

Outside more technical solutions like Free.net and Kazoo, the Internet's networked structure can be used productively to explore possibilities of alternate futures and different democracies through software endeavors: Natalie Bookchin's Agora and AgoraXchange, which bring together open source and open content, promise to pursue the (more conscious) democratic potential of the Internet. The wikipedia and other open-content projects also productively explore networking, plagiarism, and unreliability. The wikipedia, created by a multitude of users working in tandem, is the opposite of Diderot's *l'Encyclopedie* and the perverse will to knowledge explored in chapter 2, since it eschews authority: its information is not authoritative and there are no authors. The wikipedia and other such open-knowledge projects make possible the more hopeful future Jean-François Lyotard outlined in *The Postmodern Condition: A Report on Knowledge*. In this future, knowledge is not a zero-sum game because it is easily accessible to all: knowledge is not information but rather the ability to do creative things with information. Although this scenario is hopeful, we must also remember what it elides—namely, the nonplayful conflict that open-source projects generate and that constantly threatens their fate, and the myriad ways in which cooperation is *forced* on us; sharing or participating in open projects is only a start.

Thus, the Internet does not, through its town halls and chat rooms or through its disembodiment, enable publicity as imagined by the Enlightenment, nor do its protocols make its networks transparent; but it does threaten a publicity that as it makes irrelevant the distinction between public and private, enables something like democracy—a democratic potential that is constantly at risk in ideological polarizations around control and freedom.

---

58.   Ibid., 27.

ORIENTING THE FUTURE

The future remains unwritten, though not from lack of trying.

—*Bruce Sterling, Preface to* Mirrorshades

Japan is the future.

—*David Morley and Kevin Robins,* "Techno-Orientalism"

Cyberspace is a literary invention.

William Gibson first coined the term in his 1982 short story "Burning Chrome" and fleshed it out in his 1984 novel *Neuromancer* (typewritten to a sound track of late 1970s–early 1980s' punk). Preceding the conversion of the Internet into a mass medium, Gibson's Sprawl trilogy (*Neuromancer*, *Count Zero*, and *Mona Lisa Overdrive*), in conjunction with Neal Stephenson's *Snow Crash* and *Diamond Age*, would help shape computer and communications technology and ideology during the 1980s and 1990s. *Neuromancer* in particular inspired dreams of and exploits in virtual reality, mind "uploading," and e-commerce, for its console-cowboy protagonist's (Case's) description of cyberspace as a consensual hallucination dominated by *zaitbatsu* and marauding U.S. console cowboys portrayed high-speed computer networks as a commercially viable frontier of the mind.[1] In this novel, Case hustles information for money and pleasure, "liv[ing] for the bodiless exultation of cyberspace." His "elite stance involve[s] a certain

---

1.   For instance, Marc Pesce's Ono Sendai, one of the first companies working on developing Virtual Reality Markup Language, took its name from a fictional brand name in *Neuromancer*.

relaxed contempt for the flesh. The body was meat."[2] This trilogy, separated by seven-year gaps, is loosely held together by a few recurring characters and an edgy world picture, in which technology and biology have fused together, the United States of America has disappeared, zaitbatsu and *Yakuza* rule, and cyberspace, "a graphical representation of data abstracted from the banks of every computer in the human system," stands as the last frontier.

Cyberpunk's previsioning of cyberspace—or to be more accurate, Gibson's and Stephenson's previsioning of a global information matrix that has never and will never be realized, but that was nonetheless conflated with the Internet at the turn of the century (at a time when the future was reported to have arrived)—is not extraordinary for a literary genre celebrated for first imagining satellites and space shuttles.[3] Most scholars, however, view the practice of evaluating science fiction based on its predictive capabilities as apologist rather than critical. Most significantly, Fredric Jameson, himself building on Darko Suvin's influential contention that science fiction enables a subversive "*interaction of* [Brechtian] *estrangement and cognition*," has argued that science fiction uniquely defamiliarizes and restructures our experience of the present by converting it into "some future's remote past," thus enabling us to finally *experience* it.[4] Rather than simply celebrating this distilling of our present, though, Jameson contends that science fiction's inability to imagine the future reveals the "atrophy in our time of what [Herbert] Marcuse has called the *utopian imagination*, the imagination of otherness and radical difference."[5] In terms of cyberpunk, Jameson claims (in a footnote) that it is "for many of us, the supreme *literary* expression if not of postmodernism, then of late capitalism itself." Presumably, it is *the expression* because it combines "autoreferentiality"—

---

2.   William Gibson, *Neuromancer* (New York: Ace Books, 1984), 6.

3.   For a celebration of utopian science fiction and a condemnation of recent science fiction's "dark" vision, see Newt Gingrich, *To Renew America* (New York: Harper Collins, 1995).

4.   Darko Suvin, "On the Poetics of the Science Fiction Genre," *College English* 34 (1972): 375; and Fredric Jameson, "Progress versus Utopia; or, Can We Imagine the Future?" *Science Fiction Studies* 9, no. 2 (July 1982): 151–152.

5.   Jameson, "Progress," 153.

in the form of a "play with reproductive technology—film, tapes, video, computers, and the like" (which is, to Jameson, "a degraded figure of the great multinational space that remains to be cognitively mapped")—with paranoia, "the poor person's cognitive mapping in the postmodern age."[6]

Following Jameson, many critics have debated the value of cyberpunk's cognitive mapping of global capital. In particular, they have focused on whether or not its descriptions of global information networks and geographic conglomerations that dance with "biz" chart the totality of global capitalism—a totality that we cannot usually "experience." Some, such as Pam Rosenthal, argue that although cyberpunk does not offer "an adequate analysis of post-Fordist dilemmas of work and social life," it does articulate "these dilemmas in dense and intelligent ways. And its lesson is that the ideal of a final/original uncontaminated humanness is, at bottom, what is most clumsy, old-fashioned and naive about outmoded images of technological society, be they Gernsbackian, Fordist, or Marxist."[7] Others, such as Tom Moylan, assert that Gibson's fiction may insightfully map capitalism, but it produces resignation rather than subversion, for it does not offer oppositional figures and rewards individual entrepreneurship.[8] Still others, such as David Brande and Sharon Stockton, see Gibson's map itself as complicit with capitalism, since his projection of cyberspace as a frontier effectively "renders the extremely complicated flow of multinational capital both 'intelligible and commodifiable'; complexity is thus reined back to comprehensibility, and the symbolic playing field of capitalism becomes spacious again, available again for colonization."[9] The key critical debates have thus centered on the questions,

6. Fredric Jameson, *Postmodernism, or The Cultural Logic of Late Capitalism* (Durham, NC: Duke University Press, 1991), 419, 356.

7. Pam Rosenthal, "Jacked-In: Fordism, Cyberspace, and Cyberpunk," *Socialist Review* (Spring 1991): 99.

8. Tom Moylan, "Global Economy, Local Texts: Utopian/Dystopian Tension in William Gibson's Cyberpunk Trilogy," *Minnesota Review* 43–44 (1995): 54.

9. Sharon Stockton, "'The Self Regained': Cyberpunk's Retreat to the Imperium," *Contemporary Literature* 36, no. 4 (1995): 589. See also David Brande, "The Business of Cyberpunk: Symbolic Economy and Ideology in William Gibson," in *Virtual Realities and Their Dicontents*, edited by Robert Markley (Baltimore,

To what extent is cyberpunk a symptom of or diagnosis for our "present" condition, and to what extent is *Neuromancer* really postmodern?[10]

These debates remarkably assume that science fiction is always read in or tethered to the "present"—an assumption Gibson himself supports. As he explains in his 1996 interview with *Addicted to Noise*, "I'm really not in the business of inventing imaginary futures.... [W]hat I really do is look

MD: Johns Hopkins University Press, 1996), 79–106. For more on the accuracy/ worth of Gibson's cognitive mapping, see Moylan, "Global Economy, Local Texts"; Tony Fabijancic, "Space and Power: Nineteenth-Century Urban Practice and Gibson's Cyberworld," *Mosaic* 32, no. 1 (March 1999): 105–139; M. Keith Booker, "Technology, History, and the Postmodern Imagination: The Cyberpunk Fiction of William Gibson," *Arizona Quarterly* 50, no. 4 (Winter 1994): 63–87; Istvan Csicsery-Ronay Jr., "Antimancer: Cybernetics and Art in Gibson's *Count Zero*," *Science Fiction Studies* 22 (1995): 63–86; Ronald Schmitt, "Mythology and Technology: The Novels of William Gibson," *Extrapolation* 34, no. 1 (1993): 64– 78; Lance Olsen, "Virtual Termites: A Hypotextual Technomutant Explo(it)ration of William Gibson and the Electronic Beyond(s)," *Style* 29, no. 2 (Summer 1995): 287–313; John Johnston, "Computer Fictions: Narratives of Machinic Phylum," *Journal of the Fantastic in the Arts* 8, no. 4: 443–463; Inge Eriksen, "The Aesthetics of Cyberpunk," *Foundation* 53 (Fall 1991): 36–46; and Ross Farnell, "Posthuman Topologies: William Gibson's 'Architexture' in *Virtual Light* and *Idoru*," *Science Fiction Studies* 25, no. 3 (1998): 459–480. Whether or not cyberpunk is a symptom or a diagnosis—is in the end good or bad, conformist or subversive—however, is ultimately undecidable, since such a decision demands a definitive distinction between descriptive and prescriptive language, between citation and dissemination, as well as a definitive calculation of the overall message of a text based on subtracting the "bad" from the "good" strains of a text's "message."

10. See Veronica Hollinger, "Cybernetic Deconstructions: Cyberpunk and Postmodernism," *Mosaic* 23, no. 2 (Spring 1990): 29–43; Kathyne V. Lindberg, "Prosthetic Mnemonics and Prophylactic Politics: William Gibson among the Subjectivity Mechanisms," *boundary* 2 (Summer 1996): 45–83; Randy Schroeder, "Neu-Criticizing William Gibson," *Extrapolation* 35, no. 4 (1994): 330–341, and "Determinacy, Indeterminacy, and the Romantic in William Gibson," *Science Fiction Studies* 21 (1994): 155–163; Claire Sponsler, "William Gibson and the Death of Cyberpunk," in *Modes of the Fantastic*, ed. Robert A. Latham and Robert A. Collins (Westport, CT: Greenwood, 1991), and "Cyberpunk and the Dilemmas of Postmodern Narrative: The Example of William Gibson," *Contemporary Literature* 33, no. 4 (1992): 624–644; and Victoria de Zwann, "Rethinking the Slipstream: Kathy Acker Reads *Neuromancer*," *Science Fiction Studies* 24 (1997): 459–470.

at what passes for contemporary reality and select the bits that are most useful to me in terms of inducing cognitive dissonance." Specifically, Gibson sees his fiction as breaking through our ten-year buffer:

If it was 1986, we could cope. I think we have like a 10 year buffer and the buffer gets telescoped occasionally in one of those horrendous CNN moments. Like you turn on the TV and there's a building blown to shit. And it says Oklahoma City. And you can feel your brain stretch around this and the world's never going to be the same. That's now. But when we hit now, we get slammed into it like bugs on a windshield. Then we pull back and we see things are just proceeding in a normal fashion. "I can understand the world. I'm not going to freak out." I think we have to do that to survive. So I think probably what I do as an artist is I mess with that. I mess with that buffer and bring people right up close to the windshield and then pull them back and keep doing that. I suspect that's the real pleasure of the text in the sort of thing I do. I suspect that's what the people are actually paying for is having that experience. If they think they're paying for a hot ticket glimpse of the future, then they're kind of naïve.[11]

According to Gibson, his text's impact—described in visual terms—depends on its relation to its "moment" of creation (which also coincides with the moment of reading). If the work is still relevant, these moments are decontextualized as "now." Science fiction thus often has a short shelf life since it is dismissed as "misguided" once its vision of our present as its past no longer makes sense—unless it is rescued as a "classic" or conflated with ethnography (Gibson in fact dreams of being studied as a "naturalist" writer by future critics). Many sci-fi writers and critics dismiss cyberpunk as a 1980s' thing: Samuel R. Delany, for instance, argues that the destruction of technology during the Los Angeles uprising discredited Gibson's optimistic mantra, "The street finds its own uses for things."[12] Despite

---

11.   William Gibson, interview with *Addicted to Noise*, 2.10 (1996) ⟨http://www .addict.com/issues/2.10/html/hifi/Cover_Story⟩ (accessed February 1, 2000).

12.   Mark Dery, "Black to the Future: Interviews with Samuel R. Delany, Greg Tate, and Tricia Rose," *SAQ* 92, no. 4 (Fall 1993): 749.

this insistence that cyberpunk is over and done with, and Gibson's own irritation at being so "tagged," cyberpunk has continued to sell and to inspire Silicon Valley into the twenty-first century—a Silicon Valley that has consistently ignored cyberpunk's dystopian strains by conflating narrators with authors, description with prescription. Popular and academic interest in cyberpunk soared after Delany's best-before date of 1987 and Gibson's of 1994 (ten years after the publication of *Neuromancer*), propelled by the mid- to late 1990s' Internet boom. Or to be more precise, propelled by a desire to conflate *Neuromancer*'s envisioned future with our own present, propelled by a desire to see our present as the future-come-true. As such, the Internet resuscitated a text that supposedly encapsulated the 1980s' angst over post-Fordism, the Cold War, transnational corporations, and the rise of the Japanese economy. (Given the dot-bombs, Gibson will probably turn from visionary to overhyped sci-fi writer in the early twenty-first century, and those who believed or disseminated "cyberspace = the future as presently manifested" will be accused of mixing science fiction with reality. Either that, or people will begin to insist on the differences between Gibson's consensual hallucination as cyberspace versus consensual hallucination as new economy.)[13]

Debating cyberpunk's, or more often than not Gibson's, ability to express or engage the present assumes rather than examines cyberpunk's construction of past/present/future and indulges in unhelpful generalities. Debating whether or not cyberpunk's cognitive mapping supports or subverts power begs the question, What exactly makes cyberpunk a form of cognitive mapping in the first place? If cyberpunk is a form of cognitive dissonance (or following Robert Scholes's description of science fiction, if it "offers us a world clearly and radically discontinuous from the one we know, yet returns to confront that known world in some cognitive way"), what induces cognition and what estranges?[14] Importantly, cyberspace as a fiction itself relies on and constructs notions of cognition and mapping.

13. Gibson's turn to the present in *Pattern Recognition* (New York: Putnan, 2003) is an interesting symptom of this loss of the future as predictable.

14. Robert Scholes, *Structural Fabulation* (Notre Dame, IN: University of Notre Dame Press, 1975), 29.

Cyberpunk makes the invisible visible so that it can be navigated; it structurally parallels Jameson's quest to make invisible capital visible so it can be mapped. Information networks and capitalism are both invisible, and Hollywood movies from *Tron* to *Hackers* make information comprehensible through visualizations that draw parallels between humans and information bits, computer architecture and cities.[15] Therefore, rather than assume that cyberpunk offers a cognitive map, we need to analyze exactly what kind of "present" cyberpunk draws from and together—and what devices it uses to signify the past/future in order to establish "our" present as mappable. Further, given that there are little to no similarities between Gibson's matrix and the Internet, we must stop accepting cyberpunk as "originating" what we currently understand as cyberspace, and instead ask how such a conflation was accomplished and why such a conflation was/is desirable.

This chapter explores these questions through Gibson's *Neuromancer* and Mamoru Oshii's animated feature *Ghost in the Shell*, and argues that cyberpunk's tethering to the "now" stems from its high-tech Orientalism, which serves—and fails to serve—as a means of navigation. Briefly, high-tech Orientalism seeks to orient the reader to a technology-overloaded present/future (which is portrayed as belonging to Japan or other Far East countries) through the promise of readable difference, and through a conflation of information networks with an exotic urban landscape. Gibson's high-tech Orientalism has helped make his prevision of networks so influential and "originary." High-tech Orientalism offers the pleasure of exploring, the pleasure of "learning," and the pleasure of being somewhat overwhelmed, but ultimately jacked in. This pleasure usually compensates for a *lack* of mastery. High-tech Orientalism promises intimate knowledge, sexual concourse with the other, which it reduces to data or local details. It seeks to reorient—to steer the self—by making it unrepresentable and reducing everything else to images or locations (whose distances are measured temporally as well as spatially). High-tech Orientalism also enables a form of passing—invariably portrayed as the denial of a body rather than the donning of another—that relies on the other as

---

15. This also occurs within engineering itself. For instance, data is transported along "buses."

disembodied representation. This will to knowledge structures the plot of many cyberpunk novels as well as the reader's relation to the text; the reader is always learning, always trying to understand these narratives that confuse. The reader eventually emerges as a hero/ine for having figured out the landscape, for having navigated these fast-paced cyberpunk texts, since the many unrelated plots (almost) come together at the end and revelations abound. This readerly satisfaction generates desire for these vaguely dystopian futures. Thus, Silicon Valley readers are not simply "bad readers" for viewing these texts as utopian. They do not necessarily desire the future as described by these texts; rather, they long for the ultimately steerable and sexy cyberspace, which always seems within reach, even as it slips from the future to the past.[16]

High-tech Orientalism establishes information networks as a *global* (comprehensive, all-inclusive, unified, and total) navigable digital space—a conception that flies in the face of the network's current configuration. Gibson's Sprawl trilogy and *Ghost in the Shell* both compare digital landscapes to urban ones, which are exotic yet recognizable. But spatialization alone does not make cyberspace an attractive map, a desirable alternative, a place of biz; rather, the mixture of exoticism and spatialization thrills and addicts console cowboys and readers alike. Although *Neuromancer* and *Ghost in the Shell* both rely on "Far East" locations to (dis)orient their

---

16. High-tech Orientalism may be a way to steer through the future, or more properly represent the future as something that can be negotiated, but it is not simply "A western style for dominating, restructuring, and having authority over the Orient" (Edward Said, *Orientalism* [New York: Vintage Books, 1978], 3). If Said's groundbreaking interrogation of Orientalism examined it in a period of colonialism, high-tech Orientalism takes place in a period of anxiety and vulnerability. As David Morely and Kevin Robins argue in "Techno-Orientalism: Futures, Foreigners and Phobias" techno-Orientalism engages with the economic crises of the 1980s, which supposedly threatened to "emasculate" the United States and Europe (Japan became the world's largest creditor nation in 1985 and threatened to say no). Faced with a "Japanese future," high-tech Orientalism resurrects the frontier—in a virtual form—in order to open space for the United States. As opposed to the openly racist science fiction of the early to mid-twentieth century, which warned against the "yellow peril," cyberpunk fiction does not advocate white supremacy or the resurrection of a strong United States. It rather offers representations of survivors, of savvy navigators who can open closed spaces.

readers/viewers, they use different nation-states to do so, and they offer different versions of cyberspace as information map: the former portrays cyberspace as something we jack into, and the latter as something that jacks into us; *Neuromancer* takes Japan as its *Orientalis*, *Ghost in the Shell* takes Hong Kong. Both narratives, however, reorient (and hence produce) the self by turning economic threat into sexual opportunity, and although they do address the "fusion" of the technological the biological, they turn technology into biology by privileging sexual reproduction and evolution. This reorientation drives these texts' popularity and their perceived relevance to actual information technologies: if online communications threaten to submerge users in representation—if they threaten to turn users into media spectacles—high-tech Orientalism allows people to turn a blind eye to their own vulnerability and enjoy themselves while doing so. This vulnerability is economically as well as technologically induced. Both narratives were written during periods of economic duress, in which globalization seemed to equal domestic recession and loss (the 1980s for the United States and the 1990s for Japan); both portray as the site of information nations that seem to offer the greatest threats (Japan for the United States and China for Japan).

To be explicit, by reading these texts as different forms of high-tech Orientalism, I am resisting the logic that would see one, *Ghost in the Shell*, as the "native" and corrective response to the other, *Neuromancer*.[17] As Toshiya Ueno contends, the *anime* Japanoid image serves as an image machine through which "Western or other people misunderstand and fail to recognize an always illusory Japanese culture, but also is the mechanism through which Japanese misunderstand themselves."[18] This

---

17. As Rey Chow has claimed, self-representations cannot get us out of the bind of representation since they can operate as "voluntary, intimate confessions" that buttress power ("Gender and Representation," in *Feminist Consequences*, ed. Elisabeth Bronfen and Misha Kauka [New York: Columbia University Press, 2000], 43). Also, since "the self does not necessarily 'know' itself and cannot be reduced to the realm of rational cognition," and because one's experiences are not coterminous with the group one seeks to represent, self-representations do not simply correct misrepresentations (46).

18. Toshiyo Ueno, "Japanimation and Techno-Orientalism," ⟨http://www.t0.or .at/ueno/japan.htm⟩ (accessed May 1, 1999).

"misunderstanding" is itself the basis for identification—for another kind of orientation. The relation between U.S. and Japanese cyberpunk reveals a process of what George Yudice calls transculturation: a "dynamic whereby different cultural matrices impact reciprocally—though not from equal positions—on each other, not to produce a single syncretic culture but rather a heterogeneous ensemble."[19] In addition to Japanese renderings of U.S. obsessions with a Japanese future, this heterogeneous ensemble includes U.S. borrowings from anime (from the animated MTV series *Aeon Flux* to the Hollywood blockbuster *The Matrix*) and U.S. *otaku* enjoying anime such as *Lain*, which its creators Yasuyuki Ueda and Yoshitoshi Abe maintain is "a sort of cultural war against American culture and the American sense of values."[20] This complicated back and forth thus does not allow for a simple condemnation of *Neuromancer* and praise of *Ghost in the Shell* or vice versa but rather calls for a more rigorous engagement with these *global* visions. This transculturation assumes the existence of two original separate cultures, perpetuating what many scholars see as the tired and mainly rhetorical East-West division, which erases much of Asia (in order for anime to emerge as a Japanese project, many other Asian nations must be erased).[21] It magnifies what Harry Harootunian has called

---

19. George Yudice, "We Are *Not* the World," *Social Text* 31–32 (1992): 209.

20. Quoted in Kit Fox, "Interconnectivity: Three Interviews with the Staff of *Lain*," *Animerica* 7, no. 9 (October 1999): 29.

21. Just as *Ghost in the Shell* offers a vision of Japan that expands its borders, anime furthers Japan's cultural influence. This is explicit in its other name: Japanimation. Japanimation usually subsumes all animation from the Far East, obscuring the fact that much drawing is done "offshore" in South Korea. U.S.-based sites such as geocities.com further this subimperialism: in this "neighborhood"-oriented site, anime and "all things Asian" are contained within the "Tokyo" sector, effecting in virtual space Japan's past colonial ambitions. Not surprisingly, U.S. animators such as Peter Chung, creator of MTV's Liquid Television program *Aeon Flux*, resist Japanimation, insisting that "Japanese animation simply means animation done in Japan. It's not a healthy thing for people to use general terms.... It's like saying U.S. animation is all funny, talking animals" (quoted in Eleftheria Parpis, "Anime Action: Japanimation Is Edgy and Cool—and Shops Love It," *Adweek*, December 14, 1998, 20). Indeed, the terms Japanimation and

"the bilateral narcissism of the United States and Japan."[22] This bilateral narcissism, which also gets written as East versus West, makes Japan *the* representative of all things Asian or Oriental (ironically fulfilling Japan's colonial aspirations)—at a time when such a distinction does not necessarily make sense.

Desperately Seeking the Matrix

What is the matrix?

This question occupies both Gibson's readers and characters, since the matrix's "nature" changes through the Sprawl trilogy, and since Gibson's descriptions of cyberspace assume much and explain little. As mentioned in chapter 1, the well-known depiction of cyberspace as a consensual hallucination comes from an explanatory screen provided by Case's Hosaka, which Case cuts short and dismisses as a "kid's show."[23] In contrast to vague descriptions of cyberspace as comprising glowing, differently colored, and differently shaped geometric shapes (perhaps high-tech public spheres?) are copious descriptions of Case's *desire* for cyberspace. For instance, the first reference to cyberspace in the novel portrays it as an impossible dream:

A year here and he [Case] still dreamed of cyberspace, hope fading nightly. All the speed he took, all the turns he'd taken and the corners he'd cut in Night City, and still he'd see the matrix in his sleep, bright lattices of logic unfolding across that colorless void.... The Sprawl was a long strange way home over the Pacific now, and he was no console man, no cyberspace cowboy. Just another hustler, trying to make it through. But in his dreams he'd cry for it, cry in his sleep, and wake alone in the dark, curled in his capsule in some coffin

---

anime came into common parlance in the United States in order to distinguish it from the fuzzy Disney-influenced style associated with "animation" within the United States. The need to distinguish West from East thus enables a Japanification of the entire Far East.

22.   Quoted in Naoki Sakai, "'You Asians': On the Historical Role of the West and Asia Binary," *SAQ* 99, no. 4 (2000): 804.

23.   Gibson, *Neuromancer*, 52.

hotel, his hands clawed into the bedslab, temperfoam bunched between his fingers, trying to reach the console that wasn't there.[24]

By portraying Case's desire for cyberspace as sexual, Gibson naturalizes cyberspace's appeal.[25] The reference to drugs, repeated throughout the novel, also establishes cyberspace as a form of addiction so powerful that one turns to drugs to "get" over it.[26]

As the cited passages reveal, in *Neuromancer* neither the readers nor the main characters entirely "know" what is happening, although they are not entirely "lost" either. The basic plotline of *Neuromancer* is this: As punishment for stealing from one of his employers, Case is injected by the Yakuza (the mythic Japanese Mafia) with a myotoxin that makes it impossible for him to jack into cyberspace. He then travels to Night City (a subsidiary of Chiba City, Japan) in order to find a cure in its infamous nerve shops. Unable to repair the damage and out of money, Case becomes "just another hustler" on a suicidal arc. Before he manages to get himself killed, he's picked up by Molly (a female "street samurai" razorgirl/cyborg) who collects him for a mission directed by Armitage, Gibson's version of a masked man (whose standard, handsome, plastic features serve as his mask). Armitage fixes Case's nerve damage in exchange for his cooperation, and to ensure his loyalty, he lines Case's main arteries with toxin sacs. In order to prevent his nerve damage from returning, Case must be injected with an enzyme possessed by Armitage. The team first breaks into Sense/Net to steal a ROM construct (a program that mimics the mind) of Dixie (Case's now dead mentor), who will help Case break into a Tessier-Ashpool (T-A) artificial intelligence called Rio or Neuro-

---

24. Ibid., 4–5.

25. Gibson furthers this effect by reciprocally describing the sexual as cyberspatial: Case's orgasm is depicted (visually) as "flaring blue in a timeless space, a vastness like the matrix, where the faces were shredded and blown away down hurricane corridors, and her inner thighs were strong and wet against his hips." Ibid., 33.

26. Ann Weinstone, in "Welcome to the Pharmacy: Addiction, Transcendence, and Virtual Reality" (*diacritics* 27, no. 3 [1997]: 77–89), has argued that the conflation of jacking in with getting high serves as a means of transcendence.

mancer. Molly physically steals the construct while Case, jacked into her sense sensorium via simstim, staffs the virtual operation and keeps time. The real boss turns out to be Wintermute, another T-A artificial intelligence who wishes to merge with Neuromancer in order to form a sentient being: Wintermute is improvisation; Neuromancer is personality. To merge, Molly must enter Villa Straylight—the T-A's mansion in Freeside (outer space)—and extract the "word" from 3Jane (Tessier's and Ashpool's daughter), while Case hacks into Neuromancer in cyberspace with the help of a Chinese virus program. Things get complicated, but the ending is somewhat happy: Wintermute and Neuromancer merge to become the matrix; Case gets his blood changed; Molly leaves him to pursue further adventures. Throughout, Case flips between reality, cyberspace, and simstim.

As Pam Rosenthal remarks, "The future in the cyberpunk world, no matter how astonishing its technological detailing, is always shockingly recognizable—it is our world, gotten worse, gotten more uncomfortable, inhospitable, dangerous, and thrilling."[27] This thrilling danger is partly produced by the complete erasure of the noncriminal working class.[28] This shocking recognizability is produced through confusing yet decipherable references (such as "BAMA"—the Boston Atlanta Metropolitan Axis—and "the war"—the two-week World War III), gratuitous phrases, and specialized language.[29] It is also produced through visual references: in many ways, *Neuromancer* refuses the "interiority" of language and reads like an impossible screenplay rather than a novel (as mentioned in chapter 1, Gibson considers *Neuromancer* a form of nonliterary popular culture). Consider, for instance, its opening paragraphs:

27.   Rosenthal, "Jacked-In," 85.

28.   For more on the disappearance of a noncriminal working class in global narratives, see Roger Rouse, "Thinking through Transnationalism: Notes on the Cultural Politics of Class Relations in the Contemporary United States," *Public Culture* 7 (Winter 1995): 353–402.

29.   As Gibson explains in an interview, "It was the *gratuitous* moves, the odd, quirky, irrelevant details, that provided a sense of strangeness" (McCaffrey, *Storming the Reality Studio: A Casebook of Cyberpunk and Postmodern Fiction* [Durham, NC: Duke University Press, 1991], 141).

The sky above the port was the color of television, tuned to a dead channel.

"It's not like I'm using," Case heard someone say, as he shouldered his way through the crowd around the door of the Cat. "It's like my body's developed this massive drug deficiency." It was a Sprawl voice and a Sprawl joke. The Chatsubo was a bar for professional expatriates; you could drink there for a week and never hear two words in Japanese.

Ratz was tending bar, his prosthetic arm jerking monotonously as he filled a tray of glasses with draft Kirin. He saw Case and smiled, his teeth a webwork of East European steel and brown decay. Case found a place at the bar, between the unlikely tan of one of Lonny Zone's whores and the crisp naval uniform of a tall African whose cheekbones were ridged with precise rows of tribal scars. "Wage was in here early, with two joeboys," Ratz said, shoving a draft across the bar with his good hand. "Maybe some business with you, Case?"[30]

In these three opening paragraphs, Gibson matter-of-factly juxtaposes the natural and the technological, the primitive and the high-tech—all in *visual* yet jarring terms. He also uses foreign (mainly Japanese) brand names, such as Kirin, in the place of more familiar U.S. ones, such as Bud (later, he introduces odder names, such as the Mitsubishi Bank of America, Ono-Sendai, Tessier-Ashpool, Maas-Neotek). Corporate names as modifiers are essential: it is never a coffeemaker, but a "Braun coffeemaker" (and later a "Braun robot device"). *Neuromancer* also proliferates unfamiliar proper names, such as Lonny Zone and the Sprawl. These descriptors are appropriate in a series all about information: they are noninformative, but written in such a way that one thinks they should relay information. Jargon, such as "joeboy," furthers this informatic effect, for presumably such jargon makes sense to someone. This combination of jargon and foreign and made-up brand names gives the impression that this world should be knowable, or that some reader who knows should exist or emerge.

Significantly, the most important markers are racial and ethnic. Although Gibson argues that nation-states in his new world have mainly disappeared or become reconfigured, nationality or continentality (when it

30. Gibson, *Neuromancer*, 3.

comes to nonwhite characters) has become all the stronger, for geography determines type: Ratz's teeth are a webwork of East European steel and brown decay, and Case's fellow bar inhabitant is a tall African whose cheekbones are ridged with precise rows of tribal scars. The Zionites are constantly high, always touching each other and everyone else, and are generally incomprehensible.[31] Istanbul—that classically "Oriental" space—is described as a sluggish city that "never changes," seeped in history and prejudice (juxtaposing Turkey's open sexism with Molly's badass coolness makes "our" sexy technological elite appear sexism-/racism-free and makes technological enhancements seem empowering, while still adhering to a logic of the survival of the fittest). These "dark" others in *Neuromancer* are marked as technologically outside, as involved in an alternative past, a past/present/future of tribal scars and age-old ethnic hatreds, whose familiar primitivism, juxtaposed against "our future," shocks the reader. These proliferating "natives" are markers of authenticity. As Lisa Nakamura, drawing from Rey Chow's essay "Where Have All the Natives Gone?" argues, racial stereotypes serve as an auratic presence for us.[32] In an age of technical reproducibility, the never-changing native enables distance and uniqueness.[33] The constant pinning or conflation of race with location and/or time period reveals the ways in which *Neuromancer*'s global or cosmopolitan future depends on stereotypical descriptions of raced others who serve as "orienting points" for the readers and the protagonist (Case too is hardly a "complicated" character, but his

31.   Maelcum, Case's Zionite sidekick, also serves as an erotic object: Case constantly stares at Maelcum's muscular back and describes him as he would Molly.

32.   Lisa Nakamura, *Cybertypes: Race, Ethnicity, and Identity on the Internet* (New York: Routledge, 2003), 6.

33.   This denial of coevalness, as Johannes Fabian has asserted in *Time and the Other: How Anthropology Makes Its Object* (New York: Columbia University Press, 1983), is how anthropology constitutes its object: native others are treated consistently as though their existence does not take place in the same time as the ethnographer's. Cyberpunk magnifies anthropology's "time machine" effect by literally transporting the reader into the "near" future—a future made shockingly recognizable through the juxtaposition of primitive (non-Western) pasts with present (Western) ones.

character develops and surprises). Cyberpunk's much-lauded ability to make us finally "experience" our present depends on the "primitive"; this inability to move outside what Naoki Sakai has called the cartographic logic of the West versus the rest is a greater "failure" than the inability to imagine a future utopia.

Not all natives are equal: Japan plays a critical role in the cyberpunk present, for the future world "gotten worse, gotten more uncomfortable, inhospitable, dangerous, and thrilling" invariably translates into the world gotten more Japanese.[34] As Joshua La Bare claims, "The Japanese have somehow wrapped up the future, hemmed it in, taken control of it; or rather, from our perspective, Western science fiction writers have wrapped it up for them in words."[35] This Japanese future (paradoxically) depends on emblems of the Japanese past: as Lisa Nakamura notes, "Anachronistic signs of Japaneseness are made, in the conventions of cyberpunk, to signify the future rather than the past."[36] But these anachronistic signs of Japaneseness are not randomly chosen: samurais, ninjas, and shonen draw from Japan's Edo period. They confine the Japanese past to the period of first contact between the West and Japan. Cyberpunk mixes images of the mysterious yet-to-be-opened Japan (which eventually did submit to the West) with the conquering corporate Japan of the future. In addition, *Neuromancer* portrays the "near" Japanese past (that is, the present) as a technological badlands produced through contact with the West. Describing Night City, Case conjectures, "The Yakuza might be preserving the place [Night City] as a kind of historical park, a reminder of humble origins."[37] But Night City, as the opening page of

---

34. For other "Japanicized" futures, see Ridley Scott, *Blade Runner* (1982, 35 mm, 117 minutes); William Gibson, *Count Zero* (New York: Ace Books, 1986), *Mona Lisa Overdrive* (Toronto: Bantam Books, 1988), and *Idoru* (New York: G. P. Putnam's Sons, 1996); and Neal Stephenson, *Snow Crash* (New York: Bantam Books, 1992).

35. Joshua La Bare, "The Future: 'Wrapped ... in That Mysterious Japanese Way,'" *Science Fiction Studies* 17, no. 1 (March 2000): 23.

36. Nakamura, *Cybertypes*, 63.

37. Gibson, *Neuromancer*, 11.

Neuromancer makes clear, is filled with gaijin paradises, places where "you could drink ... for a week and never hear two words in Japanese." Night City—"a deliberately unsupervised playground for technology itself"— preserves the moment of fusion between East and West, the moment that the Japanese take over the development of "Western" technology. Or more pointedly, the "origin" of Japanese success is gaijin. So if as Rey Chow observes in her reading of contemporary Chinese cinema's use of the primitive, the primitive "signifies not a longing for a past and a culture that can no longer be" but wishful thinking that the primitive is the prime, *Neuromancer*'s, high-tech Orientalist primitivism does not make Japan primary.[38]

Within this grim Japanified future, cyberspace appears to be a Western frontier in which U.S. ingenuity wins over Japanese corporate assimilation, for cyberspace allows for piracy and autonomy. In stark contrast to those working for seemingly omnipotent zaitbatsu, for whom power is gained through "gradual and willing accommodation of the machine, the system, the parent organism," the meatless console cowboy stands as an individual talent.[39] Zaitbatsu, which need the console cowboy to steal data by manipulating ICEbreakers (intrusion countermeasures electronics), permit him economic autonomy (thus making him effectively zaitbatsu's dark side). The console cowboy escapes this machine-organism fusion by escaping his body—by becoming a disembodied mind—when he merges with technology, and his celebrity/success depends on his anonymity. As Pam Rosenthal argues, "The hacker mystique posits power through anonymity. One does not log on to the system through authorized paths of entry; one sneaks in, dropping through trap doors in the security program, hiding one's tracks, immune to the audit trails that were put there to make the perceiver part of the data perceived. It is a dream of recovering power and wholeness by seeing wonders and by not being seen."[40] Thus, cyberspace allows the hacker to assume the privilege of

---

38.   Rey Chow, *Primitive Passions: Visuality, Sexuality, Ethnography, and Contemporary Chinese Cinema* (New York: Columbia University Press, 1995), 37.

39.   Gibson, *Neuromancer*, 203.

40.   Rosenthal, "Jacked-In," 99.

the imperial subject—"to see without being seen."[41] This recovery of wholeness and imperialism also recovers U.S. ideals. As Frederick Buell maintains, through the console cowboy, "a cowboy on the new frontier of cyberspace, he [Gibson] brings a pre–Frederick Jackson Turner excitement into a postmodern, hyperdeveloped world; if the old frontier has been built out thoroughly and its excitements become guilty ones in the wake of contemporary multicultural/postcolonial rewritings of Western history, try, then, cyberspace in an apparently polycultural, globalized era." More succinctly, Buell claims that "cyberspace becomes the new U.S. Frontier, accessible to the privileged insider who happens to be a reconfigured version of the American pulp hero."[42]

Perhaps, but not because cyberspace is outside the Japanification of the world; cyberspace in *Neuromancer* is not a U.S. frontier, and good old American cowboys cannot survive without things Japanese. First, cowboys cannot access cyberspace without Japanese equipment (Case needs his Ono-Sendai in order to jack in). Second, cyberspace is filled with Asian trademarks and corporations; however, cyberspace—unlike the physical landscape—can be conquered and made to submit: entering cyberspace is analogous to opening up the Orient. *Neuromancer* counters U.S. anxieties about "exposure to, and penetration by, Japanese culture" through cyberspace, through a medium that enables U.S. penetration.[43] Cyberspace as disembodied representation rehearses themes of Oriental exoticism and Western penetration. Consider, for instance, the moment Case reunites with cyberspace:

> A gray disk, the color of Chiba sky.
> *Now*—
> Disk beginning to rotate, faster, becoming a sphere of paler gray.
> Expanding—

41.   Diana Fuss, *Identification Papers* (New York: Routledge, 1995), 149.

42.   Frederick Buell, "Nationalist Postnationalism: Globalist Discourse in Contemporary American Culture," *American Quarterly* 50, no. 3 (September 1998): 503, 566.

43.   Morley and Robins, "Techno-Orientalism," 139.

And it flowed, flowered for him, fluid neon origami trick, the unfolding of his distanceless home, his country, transparent 3D chessboard extending to infinity. Inner eye opening to the stepped scarlet pyramid of the Eastern Seaboard Fission Authority burning beyond the green cubes of the Mitsubishi Bank of America, and high and very far away he saw the spiral arms of military systems, forever beyond his reach.

And somewhere he was laughing, in a white-painted loft, distant fingers caressing the deck, tears of release streaking his face.[44]

Cyberspace opens up, flowers for him—a "fluid neon origami trick." Reuniting with cyberspace is sexual: he has tears of release as he enters once more his distanceless home. Molly notes, "I saw you stroking that Sendai; man, it was pornographic."[45] This flowering cyberspace draws on the same pornographic Orientalist fantasies of opening Asian beauties as mainstream cyberporn. To repeat the description of asiannudes.com cited in chapter 3:

You are welcome to our dojo! Look no further, traveler. You have found the Clan of Asian Nudes, filled with gorgeous Asian women in complete submission. Take them by becoming a samurai. Our dojo houses the most incredible supermodels from Japan, Vietnam, China, Laos, and San Francisco's Chinatown! Their authentic, divine beauty will have you entranced nightly. New girls are added almost every day, their gifts blossoming before you on the screen.[46]

Not only does cyberspace blossom for the console cowboy, so too do Oriental ICEbreakers. When Case breaks into the T-A Rio artificial intelligence Neuromancer, he uses a Chinese Kuang Grade Eleven ICEbreaker and this "big mother" "unfold[s] around them. Polychrome shadow, countless translucent layers shifting and recombining. Protean, enormous, it tower[s] above them, blotting out the void."[47] The translucent shifting

---

44.   Gibson, *Neuromancer*, 52.

45.   Ibid., 47.

46.   ⟨http://www.asiannudes.com/tour1.html⟩ (accessed April 1, 1999).

47.   Gibson, *Neuromancer*, 168.

layers surround them, evoking images of Oriental mystery and penetrability.[48] This Oriental big mother blots out the void, filling it with its shadow, revealing its secret to the Occidental male who maneuvers it to perform his will. This link between cyberspace and blossoming Oriental female positions the Western viewer as samurai, and contains the "modern" threat of Japan by remapping Japan as feudal and premodern. If, as David Morley and Kevin Robins assert, Japan "has destabilized the neat correlation between West/East and modern/premodern," this feudal portrayal reorients the Western viewer (here cowboy) by re-Orientalizing Japan.[49] Hence the allusions to the Edo and Meiji eras, which undermine the future global power of Japan.

Entering cyberspace allows one to conquer a vaguely threatening Oriental landscape. If the Yakuza—the "sons of the neon chrysanthemum"—have altered his body so that Case can no longer jack in to cyberspace, by reentering it, he takes over their territory by uniting with their flowering mother.[50] As Stephen Beard in his reading of *Blade Runner* suggests, "Through the projection of exotic (and erotic) fantasies onto this high-tech delirium, anxieties about the 'impotence' of Western culture can be, momentarily, screened out. High-tech Orientalism makes possible 'cultural amnesia, ecstatic alienation, serial self-erasure.'"[51] In *Neuromancer*, high-tech Orientalism allows one to erase one's body in orgasmic ecstasy. Or to be more precise, high-tech Orientalism allows one to *enjoy* anxieties about Western impotence. It allows one, as Gibson puts it, "to try [ ] to *come* to terms with the awe and terror inspired ... by the world in which we live" (emphasis added).[52]

Although this call to *enjoy* one's emasculation—and in this emasculated state to "jack into" another—depends on the ability to jack off and in at one's pleasure, it nevertheless offers an alternative "nerd-cool" form

---

48. For more on Orientalism and translucent layers, see David Henry Hwang, *M. Butterfly* (New York: Penguin, 1989).

49. Morley and Robins, "Techno-Orientalism," 146.

50. Gibson, *Neuromancer*, 35.

51. Quoted in Morley and Robins, "Techno-Orientalism," 154.

52. Quoted in Rosenthal, "Jacked-In," 85.

of masculinity that contrasts sharply with the Arnold Schwarzenegger type also popular in the 1980s.[53] Case is an emasculated cowboy. Although Case does save the day, Molly leaves him because happiness takes the edge off her game, and Case marries a girl named Frank. Case is often described as passive, as navigated. Feelings and insights "come to him" or "hit him," making Case the inert recipient of impulses that collide with him and that he sometimes senses earlier.[54] Case, jacking into Molly, is forced to follow her gaze and feel how tight her jeans are; when Molly and Case make love, Case is ridden. When hustling in Night City, he is "driven by a cold intensity that seems to belong to someone else." At the close of the novel, a self-loathing that makes him move "beyond ego, beyond personality, beyond awareness" fuels his victory over Neuromancer's ICE. When Molly suggests they become partners, Case replies, "I gotta lotta choice, huh?"[55] Thus, this ecstasy does not obliterate the impotence of the cowboy but rather allows him to live with it. It also reveals the limitations of such sexual fantasies and conquest, for this orgasmic ecstasy constructs cyberspace—the supposed consensual hallucination—as a solipsistic space.

In cyberspace, Case runs into no other people—or perhaps more precisely, no other disembodied minds. In the matrix, Case communicates with artificial intelligences, computer viruses, and computer constructs. These others—these codes—that Case encounters are mimics. The Chinese ICEbreaker does the methodical hacking work, going "Siamese" on the computer-defense systems. Glowing and colorful cubes in cyberspace represent Japanese corporations such as the Mitsubishi Bank of America. The closest things to sentient beings Case encounters online are Dixie (the ROM construct of his deceased hacker mentor), Linda Lee (whose

---

53.   It is as different from the pumped-up male as "sneaky fuckers" are from gorillas. "Sneaky fuckers" are male gorillas who rather than becoming silverbacks, are almost indistinguishable from females. Rather than fighting with other males over territory, they live among the female gorillas and have sex "undetected" (hence the name). As well, although nerd-cool in cyberpunk fiction is aggressively heterosexual, computer programmers are not always so.

54.   Gibson, *Neuromancer*, 36.

55.   Ibid., 7, 262, 51.

ROM construct he encounters when Neuromancer attempts to trap him), and the T-A artificial intelligences Wintermute and Neuromancer. Thus, cyberspace is "a drastic simplification" that not only limits sensual bandwidth; it also literally reduces others to code.[56]

This empty high-tech Orientalist space parallels the textual construction of the Orient in early scholarly studies that focused on ancient civilizations. These studies, as Said has argued, treated the Orient as empty; the "real" Egyptians that Orientalist scholars encountered—if these scholars traveled to Egypt at all—were treated as background relics, or as proof of the Oriental race's degeneration.[57] In cyberspace, then, as in all Orientalist spaces, there are disembodied minds, on the one hand, and disembodied representations, on the other. There are those who can reason online and those who are reduced to information. In cyberspace, there is disembodiment, and then there is disembodiment. Via high-tech Orientalism, the window of cyberspace becomes a mirror that reflects Case's mind and reduces others to background, or reflects his mind via these others. High-tech Orientalism, like its nontech version, "defines the Orient as that which can never be a subject."[58] In order to preserve the U.S. cowboy, it reinforces stereotypes of the Japanese as mechanical mimics (imitators of technology). This is not to say that in order to portray a more "fair" version of cyberspace, Gibson should have included Japanese cowboys within *Neuromancer* (or even more Japanese characters), nor is it to say that Gibson celebrates cyberspace as Orientalist. It is to say that this influential version of cyberspace mixes together frontier dreams with sexual conquest: it reveals the objectification of others to be key to the construction of any "cowboy." This is, perhaps, a brilliant critique of Orientalism in general. Perhaps.

Significantly, the Orient is first and foremost a virtual space. Said contends that the Orient is not a "real" space but rather a textual universe (that is, created by supposedly descriptive Orientalist texts). Descriptions of Case navigating both spaces make explicit the parallel between Japanese

---

56. Ibid., 55.

57. Said, *Orientalism*, 52.

58. Naoki Sakai, quoted in Morley and Robins, "Techno-Orientalism," 146.

urban space and cyberspace. When he and Molly play a cat-and-mouse game through Night City, Case says,

In some weird and very approximate way, it was like a run in the matrix. Get just wasted enough, find yourself in some desperate but strangely arbitrary kind of trouble, and it was possible to see Ninsei as a field of data, the way the matrix had once reminded him of proteins linking to distinguish cell specialties. Then you could throw yourself into a highspeed drift and skid, totally engaged but set apart from it all, and all around you the dance of biz, information interacting, data made flesh, in the mazes of the black market.[59]

When one becomes slightly disoriented (and in *Neuromancer*, Case is almost always high or in some strangely arbitrary trouble), Ninsei becomes the matrix, a world in which others are reduced to information or data. Like in cyberspace, these reductions enable a certain self-direction; they enable you to "throw yourself into a highspeed drift and skid." Parallels between cyberspace and Ninsei sprinkle *Neuromancer*. The gray disk that marks Case's entry into cyberspace is the color of the Chiba sky (the color of television tuned to a dead channel). When Case remembers Ninsei, he remembers "faces and Ninsei neon," a neon that is replicated in the bright red-and-green cyberspatial representations of corporations. Ninsei people are reduced to light and code. Case always remembers his former lover Linda Lee as "bathed in restless laser light, features reduced to a code."[60] The easy codification of things and people breaks down when Case confronts his other "home," BAMA; hence, when he is in the metropolis again and everything no longer mimics him, Case notes, "Ninsei had been a lot simpler."[61] Ninsei had been a lot simpler because this Oriental space always existed as information, as code for Case. Just as the Japanese

---

59.　Gibson, *Neuromancer*, 16.

60.　Ibid., 18. In his 1996 novel *Idoru*, Gibson takes this datafication of Asians to the extreme: Rei Teio is a virtual construct. She "grows"—that is, becomes more complicated—by absorbing information and mimicking others. People "see" her as a hologram.

61.　Gibson, *Neuromancer*, 69.

language reduces to Sony and Kirin, Ninsei as a whole—not just Chiba City—reduces to data.

Importantly, Case reveals himself to be a *bad* navigator at times. In the high-speed chase cited earlier, he correctly assesses that Molly is following him, but incorrectly assumes that she is doing so on Wage's behalf (based on misleading information given to him by Linda Lee). As Molly puts it, Case just fit her into his reality picture.[62] Linda Lee also moves from being an easily codified character to a woman (albeit as a ROM construct) who embodies the complex patterns of the human body, and although Case eventually wins in cyberspace, he flatlines several times, and Neuromancer almost seduces Case into dying there. Lastly, the neat separation between cyberspace and the physical world collapses at the end, when Wintermute's plans go astray and Case must enter the T-A villa to help Molly. In other words, the cowboy and the datafication of others do not always work; Case's rehearsing of Orientalism as a means of navigation and understanding does not always succeed. (Arguably cyberspace as a frontier and Case as a cowboy are produced through the contrast between cyberspace and simstim: simstim enables one to feel the physical sensations of another. When Case "rides" Molly via simstim, he is irritated by the fact that he cannot control Molly's gaze or her action, and he dismisses it as a "meat toy.")

Regardless, *Neuromancer* insists on the navigability and noninvasiveness of such communications, as though a consensual hallucination would not be disorienting.[63] The strains of Dashiell Hammett and a traditional

---

62. Ibid., 24.

63. Gibson does explore the ways in which this type of communication could lead to a suffocating intimacy in *Count Zero*, but only when considering artificial intelligence-to-human relations, not human-to-human. In contrast as noted later, Octavia Butler's *Patternmaster* series deals with the damaging and disconcerting effects of mind-to-mind communication through its psychotic empaths. She also emphasizes the physical costs of a consensual hallucination in her *Parable* series, in which "sharers"—children born of drug-addicted mothers—feel the pain and joy of others (more precisely, that they imagine others to be suffering). This is not to say that *Neuromancer* completely disregards the physical—cowboys can die in cyberspace, but its vision is very different from Pat Cadigan's world in *Synners*, in which a stroke advances into a network and travels from person to person.

detective story plot reinforce this navigability, this epistemophilia, this desire to seek out and understand. The narrative lures the reader along through the promise of learning: we are given more and more clues as the novel progresses so that we too can figure out the "mystery" (although not enough to figure out Neuromancer's "true name"). And this epistemophilia is tied intimately to the promise of finally getting to know the other, who is never banal and who always has a secret to be revealed (in *Count Zero*, Bobby finally visits the projects and sees what exotic secret world these concrete buildings hide).

Thus, cyberpunk's twin obsessions with cyberspace and Japan as the Orient are not accidental, and cannot be reduced to endless citations of Ridley Scott's *Blade Runner*. Rather, the Japanese Orient is a privileged example of the virtual. It orients the reader/viewer, enabling him or her to envision the world as data. This twinning sustains—barely—the dream of self-erasure and pure subjectivity. Most simply, others must be reduced to information in order for the console cowboy to emerge and penetrate. The dream of bodiless subjectivity must be accompanied by bodiless representivity. This high-tech Orientalism also renders Gibson's text something other than mere text. Through these visual spectacles and through prose that works visually, Gibson, typing to the beat of late 1970s and early 1980s' punk, responds to the disorientation around him through an imaginary Orientalist world.[64] Gibson writes what realist visual technologies could not and cannot yet represent (either as a film or a reality), and thus establishes the "originary" desire for cyberspace.

Looking Back

Cyberpunk is not simply Orientalist fiction produced to come to terms with U.S. economic "softness" and emasculation. Most significantly, cyberpunk has impacted and been impacted by a genre of anime called *mecha* (a Japanese transliteration and transformation of the word mechanical);

As well, rather than two artificial intelligences merging, a human and an artificial intelligence merge in Cadigan's *Synners*.

64.   Orientalist spectacles were also key to the emergence of film spectatorship. For examples, see Georges Méliès's Oriental trick films and the many renditions of *Ali and the Magic Lantern*.

through mecha, anime in general has gained cult status in nations such as the United States and France.[65] Cyberpunk, I argue, enables profound, compromised, obfuscatory, and hopeful identification and misrecognition between U.S. and Japanese otaku.[66] Anime's relation to cyberspace is not simply thematic: as Thomas Lamarre has observed, anime's use of limited animation makes it analogous to scanning information—to the experience of informatization.[67] As well, both cyberspace and anime offer an escape from indexicality: in both spaces, the impossible can be represented and "seen."

In the following sections I turn to Mamoru Oshii's *Ghost in the Shell* (*Kôkaku kidôtai*) in order to investigate what happens to cyberpunk when it travels *home*, so to speak, for *Ghost in the Shell* and mecha more broadly insist on the Japanese as *primary* by displacing "primitiveness" onto the Chinese. The high-tech Oriental is always in flux, always identified as the denizen of the nation-state most threatening economic and technological superiority.

Ghosts in the Shell

Ghost in the Shell, released simultaneously in Europe, the United States, and Japan in 1995 (during Japan's seemingly never-ceasing recession), was the most Westernized anime (in terms of its animation style and foreign-market outlook) produced to date. It marked anime's U.S. debut in major movie theatres (although it is still mainly aired on television in the United States). *Ghost in the Shell* reached number one on *Billboard*

---

65.   Although popular Japanese mecha series such as *Robotech* and *Astroboy* pre-date cyberpunk, mecha is now most often translated as cyberpunk, with posters for popular series such as *The Bubblegum Crisis* prominently featuring the English word cyberpunk. For the "global" popularity of mecha, see ⟨http://www.anipike.com⟩; and Laurence Lerman, "Anime Vids Get Euro-Friendly," *Variety*, June 24, 1996, 103. Also, anime directed toward a girl audience in Japan is popular among U.S. otaku.

66.   Significantly, the first cover of *Wired* magazine featured the Japanese word otaku.

67.   See Thomas Lamarre, "From Animation to *Anime*: Drawing Movements and Moving Drawings," *Japan Forum* 14, no. 2 (2002): 329–367.

magazine's video sales chart and earned the rather limited title of New York City's highest-grossing film shown exclusively on a single screen in one theater.[68] Oshii's work was a hallmark in anime production for both aesthetic and corporate reasons: *Ghost in the Shell* was "the most expensive and technically advanced Japanese animated feature yet made," although it still only cost $10 million—one-tenth of the cost of *The Hunchback of Notre Dame*.[69] It was also cofinanced and produced by Japan's Bandai and Kodansha and Chicago-based Manga Entertainment.

This anime is based on the 1989–1990 *manga* (Japanese comic book) series of the same name created by Shirow Masamune. According to Shirow, *Ghost in the Shell* is a relatively international work that transcends national boundaries, and includes multiple references to English and Japanese popular culture and literature.[70] As with *Neuromancer*, the particular type of globalization, rather than the mere fact of it, matters: in these narratives and almost all mecha, Japan is both primary and universal, donning the universalism it was forced to abandon after World War II. According to Naoki Sakai:

As the Japanese Empire expanded territorially, annexing Hokkaido, Okinawa, Taiwan, Korea, the Pacific Islands, and finally large parts of East and Southeast Asia under the umbrella of the Greater East Asian Co-Prosperity Sphere, the emperor was increasingly associated with the universalistic principle of the Japanese reign under which people of different ethnic backgrounds, of different languages and cultures, and of different residences were entitled to be integrated into the imperial nation and treated as equal subjects (the equality of which must be thoroughly scrutinized, indeed). Japan being an imperial

---

68.   Elizabeth Lazarowitz. "COLUMN ONE: Beyond 'Speed Racer': Japanese Animation Has Exploded in Popularity Worldwide; Creators of Such New-Generation Superheroes as a Female Cyber-Cop Hope to Cash in on TV Shows, Videos, and Comic Books," *Los Angeles Times*, December 3, 1996, 1.

69.   Ibid.

70.   Trish Ledoux, "Interview with Masamune Shirow," in *Anime Interviews: The First Five Years of ANIMERICA, ANIME, AND MANGA MONTHLY (1992–1997)*, ed. Trish Ledoux (San Francisco: Cadence Books, 1997), 39.

nation, the prewar emperor was rarely made to represent the unity of a partic-ular ethnicity or national culture.[71]

After the war, Sakai argues, the United States encouraged Japanese na-tionalism in order to curtail Japanese imperialism (and thus make the United States the sole source of universalism). Significantly, both the manga and anime versions of *Ghost in the Shell* depict the Japanese as glob-ally affluent and—in stark contrast to U.S. cyberpunk—militarily active. This near future is filled with high-tech tanks and weapons, and devoid of ninjas and kimonos. Without explaining how article 9 of the Japanese Constitution forbidding Japanese military buildup had been circumvented, they portray a people who have moved away from what 1990s' Japanese nationalists have called a "masochistic" or maternal-centered society.[72] This move, however, is not represented via hypermasculine Japanese male protagonists but rather (and consistently in many mecha from *The Bubblegum Crisis* to *The Dirty Pair*) through representations of strong, nonmaternal but well-endowed cyborg women. These "women" repre-sent a fantasy of equality in which women—who are not quite women—are as aggressive and puerile as men (perhaps machinic men with breasts). This representation is not unique to mecha—Molly in *Neuromancer* is a case in point. Significantly, though, these women are *protagonists* rather than sidekicks. The audience both identifies with and desires them—and this cross-gender and cross-cultural identification/desire is key to the "foreign" appeal of anime.

Although the anime and the manga both portray a globally affluent and militarily active Japan, they differ greatly in their depictions of glob-alization. The anime and the manga open with an introductory text, explaining that we are in the near future, that information networks pulse through the world, and that nation-states and ethnic groups still survive;

71. Sakai, "You Asians," 802.

72. For more on Japan as masochistic or maternal, see Marilyn Ivy, "Revenge and Recapitation in Recessionary Japan," *SAQ* 99, no. 4 (2000): 819–840; Tomiko Yoda, "The Rise and Fall of Maternal Society: Gender, Labor, and Capital in Contemporary Japan," *SAQ* 99, no. 4 (2000): 865–902; and Andrea G. Arai, "The 'Wild Child' of 1990s Japan," *SAQ* 99, no. 4 (2000): 841–863.

yet the manga makes more explicit which nations and which ethnic groups matter. The manga is set in a mythic place "on the edge of Asia, in a strange corporate conglomerate-state called 'Japan,'" whereas the anime is set in an unspecified place that resembles, and is indeed modeled after, Hong Kong (even though all the characters speak Japanese and the heads of state have Japanese surnames). The more conservative manga deals with "global" issues: from a "slave" socialist nation (presumably China) that provides the "master" nation (Japan) with cheap labor to Filipina girls who are dubbed and destroyed in order to create love dolls for the Japanese elite, from the disputed Northern Islands (which the Japanese win back from the Russians during World War IV) to Israeli manipulation of Japanese domestic politics. It refers to shifty Korean informants, ungrateful nations that demand aid in compensation for past exploitation, and robots (a prominent "ethnic" group with its own lobbying groups) that go berserk because rampant consumer capitalism discards them on a regular basis. Shirow's protagonists are profoundly antilabor: corporations and governments may be corrupt—and Section 9 (a secret intelligence agency filled with cyborgs) pursues corrupt politicians as well as terrorists—but corrupt labor and lazy workers have ruined society. In the manga's second issue, "Super Spartan," cyborg Major Motoko Kusanagi attacks a David Copperfieldesque government orphanage that illegally uses a ghost-erasing device. When a young boy approaches the Major as his savoir, the Major replies, "What do you want? Do you just want to eat and contribute nothing, to be brainwashed by media trash? To sacrifice the nation's future for your own selfishness? ... Listen, kid—You've got a ghost, and a brain ... and you can access a cyber-brain. Create your own future."[73]

The manga blames the media, in particular television, for the future's problems. The public is an annoyance: a videotape of the Major killing a boy seemingly without provocation causes an outrage that forces her to fake her own death. Driving away from an angry mob, the Major and her section head say,

---

73. Masamune Shirow, *Ghost in the Shell*, trans. Frederik Schodt and Toren Smith (Milwaukie, OR: Dark Horse Comics, 1995), 45.

Major: "If those peace activists would just deal with reality a little more effectively we wouldn't be placed in these situations."

Aramaki: "They're just like us. They hate violence ..."

Major: "They're so hypocritical. Emphasizing a lifestyle based on consumption is the *ultimate* violence against poor countries."[74]

The Major usually voices the profoundly anticonsumerist, promilitary lines in the manga (presumably, the message is more palatable coming from a woman—even one conceived and drawn by a male author—than a man).

Unlike the manga and like *Neuromancer*, the anime is set in a foreign city—this time Hong Kong. Whether or not Japan is still a nation-state is unclear, just as the status of the United States is unclear in *Neuromancer*. One can interpret this move away from overt Japanese nationalism as progressive, but such a reading ignores the importance of Orientalism to the anime. The plot of the *Ghost in the Shell* anime parallels *Neuromancer*—except that rather than an artificial intelligence seeking to be free by merging with its better half, an artificial life-form (the Puppet Master) seeks to free itself by merging with the Major. Set in Hong Kong in 2029, *Ghost in the Shell* follows the adventures of the Major, who leads Section 9 as she pursues the Puppet Master, a dangerous criminal who ghost-hacks people, inserting false memories, controlling their actions, and reducing them to puppets. The Major's entire body, or "shell," has been replaced by a titanium "Megatech Body." The human essence is encapsulated in one's "ghost," which holds one's memories.

Throughout the anime, a far more mature Major Motoko than her counterpart in the manga anxiously contemplates her humanity and hears voices, presumably her own ghost's, but as we find out later, the Puppet Master's as well. The Puppet Master and Major Motoko finally meet when the Puppet Master, lured into a buxom blond Megatech Body, is hit by a truck and brought to Section 9. That the Puppet Master has no organic brain yet contains traces of a burgeoning synthetic one disturbs the Section 9 cyborgs because it troubles the (already-compromised) dis-

---

74. Ibid., 307.

tinction between humans and machines. When Nakamura, the head of Section 6 (the diplomatic unit) and an unknown American come to claim the body, they identify it as the Puppet Master (they claim the Puppet Master is a human programmer whose ghost has been lured into a Mega-tech Body). In the middle of their explanation, the Puppet Master, who by this time is a badly mutilated blond torso, speaks the truth: the diplomatic corps hired a U.S. artificial intelligence company to develop the Puppet Master—an artificial life program—in order to assist diplomacy via espionage and other illegal activities. The Puppet Master then appeals for asylum, claiming he is a life-form: moving through the Net, he has become sentient, and since Japan has no death penalty, he cannot be terminated. In the meantime, Togusa (a Japanese and almost fully human member of Section 9) has deduced that Section 6 has illegally brought with it attack personnel wearing thermoneutic camouflage. Just as the Puppet Master pleads for diplomatic immunity, Section 6 attacks and steals the Puppet Master, with Section 9 in hot pursuit. When Major Motoko—alone—finally catches up with the men who have taken the Puppet Master, she too becomes a mutilated torso. Her close comrade/inferior officer, Batou, saves her from complete annihilation, and at her request, connects her and the Puppet Master. During this "dive," the Puppet Master takes over the Major's body and proposes that they merge. By merging, the Puppet Master can achieve death and diversify his program—he will live on through their offspring; Major Motoko can break through the boundaries that limit her as a person and access the vast expanse of the Net, which their offspring will populate. They merge just before Section 6 planes destroy the Puppet Master. Major Motoko survives, and Batou transplants her newly merged brain into a little girl's body. The anime ends with her leaving Batou's "safe house" to explore the expanse of the Net before her. Although she asks herself, "Where shall I go now? / The net is vast and limitless," the "camera" pans through the landscape of Hong Kong (see figures 4.1–4.3).

According to Toshiyo Ueno, "the choice of Hong Kong represents an unconscious criticism of Japan's role as sub-empire: by choosing Hong Kong as the setting of this film, and trying to visualize the information net and capitalism, the director of this film, Oshii Mamoru, unconsciously tried to criticize the sub-imperialism of Japan (and other Asian

| Figure 4.1 |
The "new Major" leaving Batou's safe house

| Figure 4.2 |
The Major overlooking the city

| Figure 4.3 |
The last frame: Hong Kong as a vast Net

nations)."[75] Rather than signaling an unconscious critique of Japan, how-ever, the choice of Hong Kong Orientalizes, representing the world as data. Faced with the task of representing invisible networks of informa-tion, Oshii chose a location he believed easily reduces to information:

In "Ghost in the Shell," I wanted to create a present flooded with informa-tion, and it [Japan's multilayered world] wouldn't have lent itself to that. For this reason, I thought of using exoticism as an approach to a city of the future. In other words, I believe that a basic feeling people get perhaps when imagin-ing a city of the near future is that while there is an element of the unknown, standing there they'll get used to this feeling of being an alien. Therefore, when I went to look for locations in Hong Kong, I felt that this was it. A city without past or future. Just a flood of information.[76]

As the last anime sequence reveals, rather than inherently having no past or no future, Hong Kong's landscape is *made* into a flood of information in order to represent the vast expanse of the Net. In order to "explain" cyberspace, the city becomes data, and in order to function as data, the city must be unknown yet readable. The "basic feeling" Hong Kong delivers, then, is the tourist's oriented disorientation, for tourists, not res-idents, *stand* in a public space in order to get used to the feeling of being alien. In other words, it is not simply that Tokyo is more multilayered than Hong Kong but rather that Oshii's Japanese audience is too familiar with Tokyo to be adequately disoriented. What city to the tourist, after all, is not a flood of information? By this, I do not mean to imply that all cities are alike; indeed, some are more "disorienting" than others. Yet the tourist's attempts to navigate reveal both the necessity and the inadequacy of maps.

In order to effect this familiar alienation, Oshii relies on street signs: "I thought that I could express networks which are invisible to all through drawing not electronic images but a most primitive low-tech group of

---

75.　Ueno, "Japanimation."

76.　"Interview with Mamoru Oshii," *ALLES*, 〈http://www.express.co.jp/ALLES/6/oshii2.html〉 (accessed May 1, 1999).

| Figure 4.4 |
Street scene in the first chase scene

| Figure 4.5 |
Hong Kong signs in the extended musical interlude

| Figure 4.6 |
Signs in English as well as Chinese characters

signboards piled like a mountain, that this would work well in drawing a world being submerged under information, in which people live like insects."[77] *Ghost in the Shell* relentlessly focuses on street signs that function as literal signposts for the foreign audience (see figures 4.4–4.6). Oshii glosses over the historical reasons for this informatic functioning: a Japanese audience can read these Chinese and English signs, even if they still look foreign, because of historical connections between East Asian countries via Confucian study and modernization.[78] Also, Oshii (paradoxically) juxtaposes the primitive and the modern in order to make Hong Kong a city without a history. As in *Blade Runner*, scenes of Oriental "teeming markets" punctuate *Ghost in the Shell*, and just as Gibson mixes together Edo images with high-tech equipment, Oshii mixes together traditional Chinese hats with high-tech office towers (see figures 4.7 and 4.8).[79] The Chinese "eternal present as past" serves as a low-tech future that orients the viewer to this high-tech one.

---

77.   Ibid. Ridley Scott previously used this technique in *Blade Runner*. There are numerous citations of *Blade Runner* in *Ghost*. For instance, the long musical scene in which the Major tours Hong Kong ends with manikins similar to those that appear in *Blade Runner* when Deckard tracks down the snake-stripping replicant Zhora (see figure 4.28).

78.   Ackbar Abbas argues that signs have the opposite effect on Hong Kong city dwellers: "Bilingual, neon-lit advertisement signs are not only almost everywhere; their often ingenious construction for maximum visibility deserves an architectural monograph in itself. The result of all this insistence is a turning off of the visual. As people in metropolitan centers tend to avoid eye contact with one another, so they now tend also to avoid eye contact with the city" (*Hong Kong: Culture and the Politics of Disappearance* [Minneapolis: University of Minnesota Press, 1997], 76).

79.   Japanese anime often use the Chinese to signify low-tech in a high-tech future. They feature a trip into "Chinatown," or Chinese tearooms that are marked as inferior or perpetrating bad employment practices. In *The Bubblegum Crisis* series, for example, the two women bond over a trip to Chinatown. In the prequel to *The Bubblegum Crisis*, the *AD Police Files*, bad labor practices at a Chinese tearoom marks the onset of a crisis with boomers. Ranma 1/2 turns into a girl when splashed with cold water. The female Ranma has red hair, and the male Ranma has black hair. As Annalee Newitz contends, "Ranma is not only feminized, but also associated with China, a country invaded and occupied by Japanese imperialist forces several times during the 20th century. Ranma's 'curse' is in fact a Chinese

| Figure 4.7 |
Overhead view of the market

| Figure 4.8 |
The Hong Kong market, replete with stereotypical Chinese figures and technology

The exoticism of the near future city makes cyberspace a necessary, if visually sparse map. If at first the viewer is confused by the cyberspace views that begin the anime, the viewer soon relies on them to understand the action and the locale, as do the characters themselves. Featured prominently in the chase scenes—and in fact, only in the chase scenes—

curse, which he got during martial arts training with Genma in China. Moreover, Ranma wears his hair in a queue and his clothing is Chinese: at school, the students often refer to him as 'the one in Chinese clothing'" ("Magical Girls and Atomic Bomb Sperm: Japanese Animation in America," *Film Quarterly* 49, no. 1 [Fall 1995]: 11).

| Figure 4.9 |
Cyberspace view of a car chase

cyberspace reduces pursuit to a game of hunter and prey; it erases local particularities by translating locations into a universal video player screen (see figure 4.9). Moving from "real life" to cyberspace means moving from being inundated with information to being presented with the bare navigational details. Thus, cyberspace and the city of the near future combine differing forms of orientation/disorientation to form high-tech Orientalism; they play with both exotic dislocation and navigational desire. At the same time, Oshii's version of cyberspace reveals the paucity of such an orientation: the visual simplicity of this cyberspace implies that manageable information is poor information.

Oshii also uses the Major and the city to represent cyberspace. As he notes, "Networks are things that can't be seen with the eyes, and using computers, showing a gigantic computer, would definitely not do the trick. Showing something like a humongous mother computer would be scary."[80] In order to represent the network in a less "scary" fashion, Oshii uses a humongous mother figure (see figure 4.10). In this image, the Major's enormous mutilated form blots out the void in the same manner that the big mother virus program does in *Neuromancer*. The wires attached to her body highlight her network connections and her broken form reveals her cyborg construction. Even mutilated, her connected

80. "Interview with Mamoru Oshii."

| Figure 4.10 |

From a movie poster for *Ghost in the Shell*

form represents power: she dominates the cityscape. Her jacked-in bare body makes cyberspace sexy, and Oshii's rendering of cyberspace is both erotic and simplistic, or perhaps erotic in its simplicity.

Importantly, the choice of an "exotic" or Oriental city is neither accidental nor inconsequential. During the 1990s, as Ackbar Abbas remarks, there was a concerted and anxious effort within Hong Kong to define itself politically and culturally before the 1997 transfer of the city to China. These efforts sought to displace Hong Kong's reputation as a mere port without culture, as a city of transients or transience. According to Abbas, the imminent disappearance of Hong Kong moved its culture from a state of "reverse hallucination" (which saw Hong Kong as a desert and culture as something that always came from elsewhere) "to a culture of disappearance, whose appearance is posited on the imminence of its disappearance" —that is, to "love at last sight." This reaction was not unproblematically good, for as Abbas observes, "in making it [Hong Kong] appear, many representations in fact work to make it disappear, most perniciously through the use of old binaries like East-West differences.... Disappearance is not a matter of effacement, but of replacement and substitution, where the perceived danger is recontained through representations that are familiar and plausible." As opposed to this resurrection of the East-West binary, Abbas favors "developing techniques of disappearance that respond to, without being absorbed by, a space of disappearance" and also favors "using disappearance to deal with disappearance."[81] Paul Virilio's conception of speed, which Virilio himself theorized in reaction to digitization, drives Abbas's vision of a new Hong Kong subjectivity— one that he sees developed most fully in Hong Kong's cinema's sense of elusiveness, slipperiness, and ambivalence. Virilio argues that because tele-communications networks work at the speed of light, speed becomes as important as, if not more than, time and space. Summarizing Virilio, Abbas asserts that speed creates a "breakdown of [the] analogical in favor of the digital ... [a] preference for the pixel over analogical line, plane, solid." This disappearance of the solid and the ubiquity of fast-moving images in turn leads to a "teleconquest of appearance."[82] Hence, Abbas implicitly sees Hong Kong as a digital space, but for very different reasons and in different ways than Oshii. If, for Oshii, Hong Kong personifies in-formation, and if he parallels urban and computer infrastructure to render invisible networks visible and comprehensible, Abbas views Hong Kong's "natural" affinity with information networks as historically determined and argues it must not be responded to with Orientalizing techniques. And if *Ghost in the Shell* portrays the rampant consumerism within Hong Kong as a means to reduce people to ants and to induce identity crises for its heroine, others, such as Rey Chow, claim that we must not disparage but rather see as liberatory forms of Hong Kong culture deemed vulgar and consumerist. She calls on us to remember that most of Hong Kong's people came to it voluntarily as a way of avoiding "the violence that comes with living as 'nationals' and 'citizens' of independent countries."[83]

Regardless of the disagreement between Abbas or Chow over the value of transience, the erasure of Hong Kong and its folding into Japan resonates with Japanese imperialism. To put it bluntly, "turning Japanese"

---

81.   Abbas, *Hong Kong*, 7, 23, 8.

82.   Virilio quoted in ibid., 9.

83.   Rey Chow, *Ethics after Idealism: Theory, Culture, Ethnicity, Reading* (Bloom-ington: Indiana University Press, 1998), 186.

is hardly an answer to Hong Kong's turnover to China from Britain, and Japan's interest in Hong Kong is hardly accidental. In 1995, the four tigers seemed fit—and China set to become the nation with the largest gross domestic product in the world—whereas the Japanese economy was stuck in a seemingly permanent recession. Both *Neuromancer* and *Ghost in the Shell* therefore turn to old imperialist dreams and tropes in order to deal with and enjoy vulnerability.

Turning Japanese

Although both *Neuromancer* and *Ghost in the Shell* create an "East" in order to create cyberspace, *Ghost in the Shell* does not mark Japan as West. Anime enables neither a simple Japanization of its audience nor a simple rejection of high-tech Orientalism. Rather, anime's cyberpunk propagates images that Ueno calls "Japanoid" since they are "not actually Japanese … [and exist] neither inside nor outside Japan." According to Ueno, the stereotyped Japanoid image

functions as the surface or rather the interface controlling the relation between Japan and the other. Techno-Orientalism is a kind of mirror stage or an image machine whose effect influences Japanese as well as other people. This mirror in fact is a semi-transparent or two-way mirror. It is through this mirror stage and its cultural apparatus that Western or other people misunderstand and fail to recognize an always illusory Japanese culture, but it also is the mechanism through which Japanese misunderstand themselves.[84]

If U.S. cyberpunk reduces the Japanese to mimics to serve as mirror images for their protagonists, anime makes this mirror two-way—on the other side, the Japanese (or at least the Japanese otaku) similarly identify with and misrecognize themselves through this image. That is, the corrupted Japanoid images circulating in the United States (the Japanese as lacking individuality and as producing ideal family units; as perversely enjoying work and as ideal workers) recirculate to Japan and affect Japanese representations. If U.S. cyberpunk makes the future Japanese in order to effect cognitive dissonance, however—to register a "future

84. Ueno, "Japanimation."

gotten worse, gotten more uncomfortable, inhospitable, dangerous, and thrilling"—cyberpunk anime perpetuate Japanoid images in order to preserve Japan as primary, and also place the blame for the future's problems back U.S. multinationals.

The portrayal of race in anime often confuses U.S. (and other) viewers, who read the anime protagonists' enormous eyes and seemingly fluid racial features as imagining a happy U.S.-style multicultural future (along the lines of "Anthem"), or as representing a Westernization of Japanese beauty standards.[85] To view anime as multicultural, though, one must reduce multiculturalism to minorities acting like the majority, for anime portrays a future world in which everyone has turned Japanese. As Annalee Newitz argues:

What these *anime* act out is a fantasy in which people of all races and Japanese people are interchangeable.... While this kind of ideology might seem satisfying and "right" to Americans raised in a multiculture, we must also remember that the Japanese are not multicultural. The ideological implications of these representations are more complex than something like "racial harmony." This multicultural fantasy takes place largely in Japan and all the races are speaking and being Japanese.... In a way, the *anime* want to imply that Americans are Japanese. If Americans are already Japanese, then it should be no surprise to any American that Japan, economically speaking, already owns a large portion of the United States.[86]

According to Newitz, anime's multicultural "cast" appropriates U.S. multiculturalism so as to naturalize Japanese economic dominance (this assumes that U.S. multiculturalism does not dream of everyone acting alike and speaking English). To make this argument, Newitz ignores the difference between the interchangeability of racial features and the interchangeability of races. Infrequent anime viewers may only recognize

---

85.   See Mark Binelli, "Large Eyes Blazing, Anime Offers Exotic Views," *Atlanta Constitution*, October 27, 1995, P10; and Jonathan Romney, "Manga for All Seasons: A Festival at the NFT Shows There Is More to Japan's Cult Anime Movies Than Misogyny and Apocalyptic Animation," *Guardian*, May 4, 1995, T.015.

86.   Newitz, "Magical Girls," 13.

| Figure 4.11 |
Japanese console operator and Chief Aramaki, head of Section 6

Japanese characters through their surnames, but these big-eyed characters usually hold positions of power. Although *Ghost in the Shell* features Japanese characters with smaller eyes than other anime, its Japanese characters do not look stereotypically Japanese, with the exception of the Chief, whose wisdom is marked by an almost Confucian countenance (see figures 4.11–4.14).[87] That Japanese anime and manga characters do not look stereotypically Japanese (the question being, Stereotypical to whom?) is not

---

87.   His appearance in the anime is a marked improvement over his appearance in the manga. In the manga, he is given apelike facial features and referred to as "ape face." More experienced anime viewers herald *Ghost in the Shell* as "a watershed in anime character design. The figures are drawn with truer anatomy: the heroine no longer has a 12-year-old's face and a pair of double D's and long legs. Her body (naked, of course, because, uh, her camouflage can't work with clothing) is rendered proportionally accurate with realistic body movement, as evidenced in a scene where she maneuvers a perfect roundhouse kick—crack—into the face of her opponent" (Edmund Lee, "Anime of the People," *Village Voice*, April 9, 1996, 15). First-time anime viewers' impressions of *Ghost in the Shell*, however, reveal the comparative nature of "truer anatomy." Laura Evenson, for instance, describes Major Motoko as sporting "the body of a Baywatch babe, the face of a beauty queen and the soul of a machine" Evenson ("Cyberbabe Takes on Tokyo in 'Ghost'; Tough, Topless Cartoon Heroine," *San Francisco Chronicle*, April 12, 1996, D3). The female figures in particular retain the large eyes prevalent in portrayals of both genders in other anime.

| Figure 4.12 |
Major Motoko Kusanagi

| Figure 4.13 |
Officer Togusa, noncyborg member of Section 9

| Figure 4.14 |
Ishikawa, member of Section 9

surprising. All animations produce images that are not indexical. Nonrealist drawings serve as the basis for most animation, and America's most famous animated big-eyed character, Mickey Mouse, certainly does not "represent" Americans indexically, even though he does as a trademark. The manner of distortion does matter, though, and the enormous eyes stem from post–World War II Japan. According to Mary Grigsby, "Before the Japanese came into contact with westerners, they drew themselves with Asian features. After contact with the west, particularly after World War II and the subsequent reconstruction of Japan under the domination of the United States, they began to depict characters that are supposed to be Japanese with western idealized physical characteristics: round eyes, blond, red or brown hair, long legs and thin bodies."[88]

These "Western" features, however, do not simply reflect "Western" beauty standards. Although these new bodies are not stereotypically "Japanese," they certainly are not "Western" either: "Westerners" (and I presume by this phrase Grigsby means white people) no more resemble these characters than do the Japanese (although rampant plastic surgery and hair coloring makes the resemblance to the Japanese more compelling). The enormous eye size arguably parodies the difference between so-called Westerners and Japanese, producing new images that would defy racial categorization, if they did not represent Japaneseness: even though Japanese characters may look less visibly Asian, Chinese characters are portrayed in a manner reminiscent of "yellow peril" propaganda (see figures 4.15 and 4.16). U.S. males are given a more "realistic" portrayal when they are marked as "Americans" as opposed to Japanified Americans. The U.S. programmer in *Ghost in the Shell*, for instance, has smaller eyes than Togusa and a large protruding nose (see figure 4.17). (The visual distinctiveness of the Japanese occurs in manga as well as anime. Frederik Schodt maintains that "in the topsy-turvy world of Japanese manga, although Japanese characters are frequently drawn with Caucasian features, when real Caucasians appear in manga they are sometimes shown as big hairy brutes. Chinese and Korean characters are frequently drawn with slant eyes and

88.   Mary Grigsby, "Sailormoon: Manga (Comics) and Anime (Cartoon) Super-heroine Meets Barbie; Global Entertainment Commodity Comes to the United States," *Journal of Popular Culture* 32, no. 1 (Summer 1998): 69.

| Figure 4.15 |
Kwan, puppet manipulated by the Puppet Master

| Figure 4.16 |
The "bad guys" who steal the Puppet Master's body

| Figure 4.17 |
U.S. artificial intelligence expert Dr. Willis

buckteeth, in much the same stereotyped fashion Japanese were depicted by American propagandists in World War II.")[89] Thus, these exaggerations, rather than making race more fluid, reinscribe racial difference.

Visual differences between those marked as Japanese versus Chinese and between those marked as Japanese versus American separates Japan from both the West and the East, making the Japanese singular and primary. These visual differences remind us that "in this cultural climate, a Japan imaginarily separated from both West and East is reproduced again and again in the political unconscious of Japanimation (subculture)."[90] As in *Neuromancer*, "others" must be conspicuously marked in order for the self to emerge as unmarked. As mentioned earlier, the popularity of Walt Disney is linked to Mickey Mouse and its cast of *animal* characters that can travel across cultures without being necessarily identified as American, while at the same time being heavily identified as such. Arguably, the large Japanese eyes are a citation of Mickey Mouse—or at the very least, an attempt at racial obscuring that makes the Japanese-named characters universal. Indeed, animation generally enables a cross-cultural exchange that exceeds the logic of same-based identification. The popular characters are toys and animals, and recent story lines focus on cross-cultural events and alliances, from *Pocohantas* to *Mulan*. When watching anime, one is free to identify with characters one would normally not: with mice and men, with women and toys. Animation structurally parallels (myths of) cyberspace, since both these "spaces" suspend indexicality and are thus spaces in which race need not matter, and yet does profoundly. As Sergei Eisenstein has argued, animation carries with it a certain omnipotence.[91]

---

89.   Frederik Schodt, *Dreamland Japan: Writings on Modern Manga* (Berkeley, CA: Stone Bridge Press, 1996), 66.

90.   Ueno, "Japanimation." Yet the only putatively U.S. female, the Puppet Master in a female body, is similarly given enormous eyes and breasts, unlike other anime such as *AD Police Files* that give U.S. women such as Caroline Evers smaller eyes and a taller physique. The similarities between the Major and the Puppet Master may stem from the fact that they both inhabit Megatech Bodies, which seem to come in two versions: blond and black haired.

91.   Sergei Eisenstein, *Eisenstein on Disney*, trans. and ed. Jay Leyda (London: Methuen, 1988).

Further, in *Ghost in the Shell*, Japaneseness becomes humanness. The only named human (noncyborg) characters are Togusa, Chief Aramaki, and the director of Section 6, Nakamura. An interchange between Togusa and Major Motoko reveals the importance of humanness/Japaneseness. After the Major reprimands Togusa for favoring a simple revolver over a more powerful weapon, Togusa asks her, "Why'd you ask for a guy like me to be transferred from the police?" Major Motoko responds that she recruited him precisely because he is a guy like himself, "an honest cop with a clean record. And you've got a regular family. With the exception of your cyber-net implants, your brain is real. No matter how powerful we may be fighting-wise, a system where all the parts react the same way is a system with a fatal flaw. Like individual, like organization. Overspecialization leads to death. That's all." Togusa's difference is his/"our" humanness, his regularity and banality, and in the end, Togusa's humanness saves the day: he figures out that Nakamura and Dr. Willis have brought in thermo-camouflaged fighters, and he uses his revolver to plant a tracking device into the escape car. Diversity, then, moves from racial diversity to diversity between cyborgs and humans, where humans who are recruited or needed for "good" diversity are Japanese.

As humanness is mapped onto the Japanese, technology and global multinational corporations are mapped back onto the United States. In a move that reverses *Neuromancer*'s dissemination of Japanese trademarks, *Ghost in the Shell* marks technology—specifically computer technology—as American through loan words. If Sony stands in for monitors in general in *Neuromancer*, transliterated words such as hacking, programmer, debug, kill, and virus brand computer technology as American (although not corporate). Anime itself is a transliteration of animation and heavily influenced by Disney. U.S. corporations such as Megatech and Genotech are the source of irresponsible capitalism, as opposed to zaitbatsu. The Puppet Master is initially believed to be American and was developed by the American company Neutron Corp. Given Japan's relation to technology, from modernization initiated during the Meiji period to the atomic bomb (a history made clear by Neutron Corp.), this insistence on technology as American rather than Japanese makes sense and reverses an aspect of high-tech Orientalism. In fact many manga, especially those that are mecha or *hentai* ("perverted"), have English names, so that anime marked as "perverse" by the United States (as well as by the Japanese) is marketed as American. The other name for hentai is *etchi*—which is a transliteration of the letter *h*.

But *Ghost in the Shell* cannot be reduced to "West equals technology" and "Japanese equals human"; the word ghost reveals the Japanese reworking of U.S. technology—and indeed U.S. culture. This word, which encapsulates the essence of a human being (like a soul but not quite), is a loan word. Since there are many preexisting Japanese words to describe one's soul or spirit whereas there are not for words such as programmer, ghost belies the usual use of loan words. Given Shirow's knowledge of U.S. technology and literature, ghost probably refers to ghost in the machine. Rather than simply alluding to Western theories of ghosts and machines, however, ghost encapsulates the forms of identification, appropriation, and transference involved in anime's reworking of high-tech Orientalism. As Diana Fuss contends, "Identification ... invokes phantoms. By incorporating the spectral remains of the dearly departed love-object, the subject vampiristically comes to life."[92] *Ghost* marks the vampiristic creation of the Japanoid subject, a subject that exceeds identification with its object and also exceeds the object itself. *Ghosts* result from an incorporation of and desire for technology. Further, only after one has imbibed technology does one's former self become an "original body"—technology thus both makes a retreat to a pure "Japanese" self impossible and enables the notion of a pure self to emerge in the first place.

The question of a ghost and its relation to humanity drives *Ghost in the Shell*. Major Motoko questions her humanity, a humanity unanchored from her "original" human form, for a good portion of it. Images that trace the Major's bodily creation from exoskeleton to womblike fleshification punctuate the opening credits (see figures 4.18–4.23). As to be expected from a film directed at adolescent boys, these images focus on her naked form, especially her breasts. Nevertheless, in contrast to the initial naked image of the Major diving from a building, this animation series emphasizes her difference from "normal" naked women by revealing her construction. Her exoskeleton is unattractive; her breasts are exposed, but as scales fall off her flesh. This extended sequence, interspersed with

---

92.   Fuss, *Identification Papers*, 1.

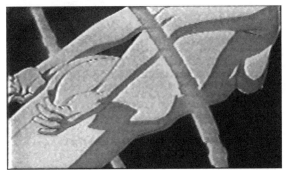

| Figure 4.18 |
The beginning exoskeleton

| Figure 4.19 |
Adding flesh under water

| Figure 4.20 |
Breaking water and emerging as flesh

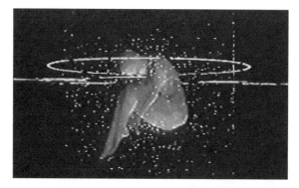

| Figure 4.21 |
The full human in the fetal position mapped in cyberspace

| Figure 4.22 |
The finished product

| Figure 4.23 |
Looking at her hand

| Figure 4.24 |
Looking back at the translator

credits that appear first as a series of ones and zeros before emerging into words (one among many features that *The Matrix* would copy), visualizes her creation and provides a way of understanding the Major as a cyborg.

This explanation, though, poses more questions than it answers, and the concluding section (the Major wakes up, looking disconnectedly at her hand and then at the Hong Kong landscape) shows the Major similarly dissatisfied. Moving from her hand to the expanse of the Hong Kong landscape, she appears to be trying to understand where her body ends and begins. She resembles an infant, using images around her to understand the relation between her parts and her whole. Throughout the anime, the Major looks for resemblances or images (and is caught by them): the anime shows her looking longingly at other cyborgs and also shows her look being arrested by theirs. When the Foreign Minister's translator is "ghost hacked" and the Major is ordered to pursue the Puppet Master, the Major pauses, looks over her shoulder, and her (and indeed our) look rests on the translator's face (figures 4.24 and 4.25). The buxom blond translator, lying on the couch with her shirt unbuttoned and her body connected to the network, is the object of the "camera's" and the Major's scopic desire. And yet, this shot also troubles the line between desire and identification (which of course is never clear), for the cables protruding from the translator's head are similar to the Major's and the scene begins with a green cyberspace view of the translator's brain—a view identical to the "read" of the Major's brain during the

| Figure 4.25 |
The ghost-hacked translator

| Figure 4.26 |
The office lady

creation sequence (importantly, only "women" are shown jacked in). The prolonged sequence in which the Major travels through Hong Kong repeats this mirror effect between the Major and cyborg others. Although her look is first arrested by an "office girl" resembling herself, it ends on manikins in an office tower (see figures 4.26–4.28).

Major Motoko is undergoing a second mirror stage—a mirror stage that will inaugurate a new subject that is neither human nor machine/ computer. Estranged from her body and faced with its lack of physical uniqueness, she searches for a way to emerge as a unique cyborg subject. The Major makes explicit her anxieties after she and the Puppet Master's mutilated form first exchange looks. In the elevator with Batou, she asks him:

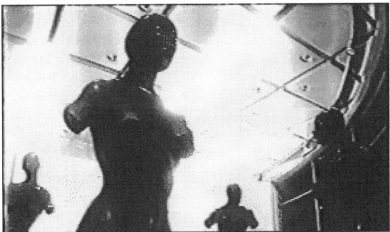

| Figure 4.27 |
The Major looking up at the office lady

| Figure 4.28 |
Manikins on display

Major: Doesn't that cyborg body look like me?

Batou: No, it doesn't.

Major: Not in the face or the figure.

Batou: What then?

Major: Maybe all full replacement cyborgs like me start wondering this. That perhaps the real me died a long time ago … and I'm a replicant made with a cyborg body and a computer brain. Or maybe there never was a real "me" to begin with.

Confronted with mirrors around her, the Major realizes what others repress: namely, that misrecognition grounds identity, that the ego is fundamentally an alter ego—that "there never was a real 'me' to begin with." This second mirror stage, however, is based on resemblances invisible to the naked eye. Having incorporated technology into herself, the Major identifies with it whenever she sees artificial forms. The scene in which the Major monitors the exchange between Sections 6 and 9 reveals this second mirror stage most explicitly (see figure 4.29). In this "shot," the "camera" mediates and indeed produces the mirroring effect between the Major and the Puppet Master. Thus, if the Major is to make sense of her body as a whole, she will have to do so through the very technology that has provoked her crisis. The viewer looks over the Major's shoulder, which also suggests that the Puppet Master mirrors the viewer.

| Figure 4.29 |
The Major and the Puppet Master

Not unexpectedly, the Major resolves her crisis—she moves from childhood toward adulthood—by merging with the Puppet Master. Before they do so, the Major repeats Togusa's earlier question by asking the Puppet Master, "Why did you choose me?" S/he replies, "Because in you I see myself" (see figures 4.30 and 4.31). At this point, these statements are literally true (when the Major dives into the Puppet Master's body, the Puppet Master takes over the Major's body), and the Puppet Master's

One last question:
Why did you choose me?

Because in you I see myself.

| Figure 4.30 |
The Major (as the Puppet Master) asking, "Why did you choose me?"

| Figure 4.31 |
The Puppet Master as the Major responding

response reveals that rather than the Major simply seeking out images or ghosts, ghosts have been pursuing her. The Puppet Master tells her, "At last I'm able to channel into you. / I've invested a lot of time into you." The voice that the Major has been listening to, and assuming was her own, is the Puppet Master's; the ghost in the shell is not her own soul but the Puppet Master's. Their merging incorporates within the Major the voice that she has been unable to hold without. It also fulfills a fantasy: the Major has been recognized—she is the object of the other's desire. When they merge, they are transplanted into a child's form, which paradoxically represents maturity; the new being replies to Batou's question (Who are you?) by finishing a biblical quotation that the Puppet Master whispered to the Major earlier: "When I was a child, I spake as a child / I understood as a child, I thought as a child. / But when I became a man, I put away childish things. / For now we see through a mirror darkly." In this passage, taken from 1 Corinthians 13, Paul ruminates about love and incorporation with God, representing this incorporation as the ability to *see through* the mirror, so that mirrors no longer reflect images but enable a vision, however dark, to an outside.[93] Through this merging, s/he has become a man—or rather a cyborg (in the manga, rather than being transplanted into a female child's body, she is put in a transvestite's one). Joining the vast Net, she has finally been able to move from part to whole by paradoxically dispersing herself.

One can read this merging as an allegory for the Japanese adaptation of U.S. technology and Japan's surpassing of the United States via this technology. As they fuse, the Puppet Master—the U.S. artificial life—dies, and the Major survives as something different, but the Major's recognizably Japanese child's body represents a Japanese future. Since the Puppet Master and the Major are represented as "love interests," the incorporation of the other seems to follow the psychoanalytic model of desire and identification. The lost love object becomes incorporated into the self in order to survive—that which cannot be held outside gets incorpo-

---

93.   As Marc Steinberg has argued, the presence of Christian references in *Ghost in the Shell* and other anime is a form of internationalism (personal correspondence). Less than 1 percent of the Japanese population is Christian.

rated within. If we map the Puppet Master as U.S. technology, then it becomes incorporated within the Japanese self.[94]

This privileging of sexual reproduction, as well as the references to psychoanalysis, philosophy, religion, and other originary myths, calls into question Donna Haraway's utopian claim that cyborgs lie outside sexual reproduction, psychoanalysis, religion, and so on. According to Haraway, cyborgs replicate rather than reproduce: "Modern medicine is also full of cyborgs, of couplings between organism and machine, each conceived as coded devices, in an intimacy and with a power that was not generated in the history of sexuality. Cyborg 'sex' restores some of the lovely replicative baroque of ferns and invertebrates (such nice organic-prophylactics against heterosexism). Cyborg replication is uncoupled from organic re-production."[95] Rather than celebrating replication, both *Neuromancer* and *Ghost in the Shell* are driven by an urge to merge that privileges sexual reproduction. In *Neuromancer*, Wintermute is driven by an urge—pro-grammed into him by Marie-Claire Tessier—to merge with his unknow-able other, Neuromancer. Such a unification, which happens at the end, produces an entirely new life-form that is, momentarily, the matrix. This new life-form then meets with a mate from Alpha Centauri and inexplica-bly shatters into many "children," who live in the matrix and who humans treat as spirits or loas. In *Ghost in the Shell*, when asked why "he" does not simply copy himself, the Puppet Master replies, "Copy is copy." Looking up at the "tree of life," the Puppet Master makes it clear that humans have summitted the tree through sexual reproduction (recombinant DNA). This denigrates most life-forms; it also obfuscates the fact that many parts of the human being reproduce asexually, and that copying errors introduce diversity (acknowledging this constitutive error would also mean acknowl-edging technology failures).

---

94. This theme is repeated in the *AD Police Files* series in which the Japanese policewoman replaces her organic eye with a mechanical one after the American cyborg/ripper Caroline Evers dies a violent death. The ghost in the shell is an identification with U.S. technology—one that lives on in a Japanese body.

95. Donna Haraway, *Simians, Cyborgs, and Women: The Reinvention of Nature* (New York: Routledge, 1991), 150.

While these narratives favor sexual reproduction and seek to insert themselves within a history of human sexuality, they also rewrite the significance—and even the means of—sexual reproduction. Recombination is favored over replication, but both "parents" die in this merging, which is by no means heterosexual. In *Neuromancer*, the two artificial intelligences are referred to as "he," and in *Ghost*, the Major and the Puppet Master are both female forms when they merge, although the Puppet Master is called a "he" and speaks in a booming male Japanese voice. This merging in and of itself is hardly subversive ("lesbian" after all can be a heterosexual pornographic category), but it loosens gene recombination from heterosexual intercourse. This fictional loosening also parallels reality, for scientists are working to produce a gamete from two *X*s. Sexual reproduction is thus becoming an effect of—or one possible route toward—gene recombination, rather than its source, and this "dissemination" of sexual reproduction complicates the status of the sexual, opening new forms of sexuality.

Who's Zooming Who?

Although *Ghost in the Shell* can be interpreted as an allegory for incorporation, such an interpretation reduces the Major to Japan and the Japanese viewer to the Major, and fails to account for the popularity of this anime within both Japan and the United States, especially within male "minority" cultures.[96] Moreover, the fact that the Major and the Puppet Master are both females when they merge belies a simple nationalist reading, since most nationalist allegories are unforgivingly heterosexual. One could account for the presence of female cyborgs and "lesbian sex" through the tradition of Japanese manga: same-sex relationships have been featured in manga from the Edo period, and *Bishonen* (pretty boy) manga feature male homosexual relationships.[97] In addition, anime such as *Ranma 1/2* and

---

96.   Scott Mauriello, one of the owners of Anime Crash, a hangout for anime fans, notes, "Anime is especially popular with minorities.... All the stories talk about a small group fighting against the system" (quoted in Lee, "Anime").

97.   See Sandra Buckley, "'Penguin in Bondage': A Graphic Tale of Japanese Comic Books," in *Technoculture*, eds. Andrew Ross and Constance Penley (Minne-

Birdy portray boys whose bodies become female under certain circumstances. Both *Bishonen* and *Ranma 1/2*, however, are not meant to realistically portray gay relationships or male transsexuals. They are not written for or by gay males but rather by women for young girls (given that anime's mainly male overseas audience does not tend to distinguish between genres, *Ranma 1/2* and *Birdy* are more popular among boys than girls in the United States).[98] So why the persistence of women and transgendering? According to Annalee Newitz, reproduction plays a key role: "Bodies manipulated by *mecha* science are merged with pieces of technology in order to 'give birth' to new creatures.... Female bodies and sexuality are therefore 'best suited' to *mecha*—and male bodies and sexuality are disfigured by it—precisely because it is related to reproduction and giving birth."[99] Cyborgs do, in a sense, give birth to new bodies; in *Ghost in the Shell*, the Major gives birth to offspring that populate the Net, and in *Bubblegum Crisis*, women bond with mechanical outer shells to become new creatures with extraordinary fighting powers. Still, this explanation overlooks transformations to reproduction and the connection between feminism (or perhaps more precisely postfeminism) and technology as empowerment.

Technological empowerment draws from and maps itself onto feminist empowerment. At about the same time that Haraway called for feminists to embrace technology in her "Cyborg Manifesto," technology companies were embracing "feminism." From Apple Computer's female runner in its mythic 1984 commercial to MCI's marked spokespeople, technology corporations, in order to sell technology and deflect questions

---

apolis: University of Minnesota Press, 1991). Frederik Schodt also claims that "homoerotic relationships have been a staple of girls' comics for years, starting with stories that featured cross-dressing women, then beautiful boys in boarding schools falling in love with each other, and so forth.... [N]owadays girls' comics with a gay theme sell, and those without one don't" (*Dreamland Japan*, 185).

98.   See Seiji Horibuchi, "Interview with Rumiko Takahashi," in *Anime Interviews: The First Five Years of ANIMERICA, ANIME, AND MANGA MONTHLY (1992–1997)*, ed. Trish Ledoux, 46–53.

99.   Newitz, "Magical Girls," 9.

of inequality, have perpetuated images of women who, with the proper technological enhancements, overcome physical inferiority to become fully functioning "equals." The belief that women need technological "extras" because they are naturally weaker underlies this narrative of technology as the great equalizer, and this narrative largely draws from (more critical) science fiction. In cyberpunk fiction, these enhancements often necessitate the loss of reproductive organs. The Major, after she has been technologically enhanced, cannot reproduce organically and her lack of sex organs becomes a joke: when Batou tells the Major that there is a lot of static in her brain, the Major replies that it is that time of the month.[100] The similarity between "geeky" boys and anime females does not mean, however, that the audience simply identifies with these women. Technology as empowering renders these women understandable and sympathetic, but does not adequately explain their prevalence.

Another possible answer would be: these women are sexy. The Major, in the manga and anime, combines cyborg and pinup—a combination whose genealogy Despina Kakoudaki convincingly outlines in her "Pinup and Cyborg: Exaggerated Gender and Artificial Intelligence." Examining 1940s' pinups placed next to military equipment or portrayed using the telephone, she argues, "The co-optation of the pinup into an instrument of war has great ideological repercussions: It admits women's relation to the military industrial complex and the increasing freedom it implies, but also conforms this new power into a pornographic sub-

---

100. The loss of reproductive organs via technological replacement and then their ghostly reappearance is the theme of the first two files of *AD Police Files*. The first concentrates on how replicants or Boomers with mechanical female sex organs go crazy—how passion and emotions are fused onto these entirely mechanical beings via ovaries. In the second file, an American woman, Caroline Evers, has her reproductive organs cybernetically replaced in order to get a promotion (she is denied her first one because her male competitor produces a graph that shows that her productivity falls with her period). After she has her reproductive organs replaced, her work becomes flawless and she becomes president of the company. She eventually marries her competitor, who then cheats on her with prostitutes and tells her that "real women are better." She kills him and then starts killing prostitutes whenever she gets menstrual cramps from her phantom period.

ject."[101] Basically, cyborgs have always been pinups. The female cyborg and all "new women" have always been "interpolated" with pornography —partly as a means to diffuse their transgressive potential, but also partly because such transgression is desired. This *appropriation* of the pornographic mode has also been a means by which the artificial woman has emerged (materialized) as an agent. In terms of Major Kusanagi, Kakoudaki alleges:

Ghost in the Shell depicts the artificial woman as a complex and sexual being. At the same time, the film demonstrates anxiety regarding repressing or counteracting this possible positive female image. As is the case with "No Woman Born," the cyborg narrative proposes the artificial woman as an agent. This narrative also redirects her, uses that agency to tackle a different target. Cyborg science fiction thematizes existential dilemmas, skin tropes, and narratives of emergence. The tradition and historical precedent of "New Women" who face representational and technological challenges—and the affinity of women's representational tropes to transparency and fetishism—affect the contemporary science fiction landscape. Faced with a space that may make consciousness disappear, the ability of women to "appear" is thus used as a means to escape the existential dilemmas of new technology.[102]

Kakoudaki insightfully observes that through sexuality and gender, cyborgs have mattered and that cyberpunk uses the female cyborg's appearance (or to-be-looked-at-ness) to escape existential dilemmas. But representations of the *female* look—especially within anime—also negotiate and humanize new technology. Importantly, the Major (and much pornography) enables simultaneous desire and identification.

Cinematically, the female look has been considered contaminated and incapable of adequate separation from its object. As Mary Ann Doane has pointed out in her groundbreaking *The Desire to Desire*, the female

---

101. Despina Kakoudaki, "Pinup and Cyborg: Exaggerated Gender and Artificial Intelligence," in *Future Females, the Next Generation*, ed. Marleen S. Barr (Boulder, CO: Rowman and Littlefield, 2000), 165.

102. Ibid., 186–187.

spectator has been viewed as "too close" to images (incapable of fetishistic distance because of her castration) and "too close" to commodities (because she is both commodity and consumer).[103] The classical Hollywood gaze has thus been considered male. According to Laura Mulvey, this gaze, in order to circumvent the castration anxiety provoked by the female spectacle, oscillates between fetishistic scopophilia and sadistic voyeurism.[104] Feminist scholars have often assumed that women either narcissistically identify with the woman as spectacle or "pass" as male. As Doane contends, neither works with women's films (films in which the woman's gaze is prominent).[105] For Doane, women's films—with their imagined rather than real female spectator—tend to desexualize and hence disembody the female spectator; this disembodiment is hardly empowering, since a bodiless woman cannot see. Regardless, women's films produce perturbations and contradictions within the narrative economy.

Ghost in the Shell is hardly a traditional woman's film. Although its protagonist is female, it seems closer to soft porn (without overt female sexual pleasure and without an appeal to a male spectator) than a woman's weepie. Yet *Ghost in the Shell* both pornographizes cyborgs and disseminates a thoroughly contaminated, "close," and seemingly disembodied female gaze: the look, in other words, is gendered (machine) female. The female cyborg represents ideal cyborgian subjectivity; the Major is always absorbed by the spectacles around her and incapable of distinguishing herself from others. As well, her more than evident castration denies male spectators the fetishism needed to separate themselves from the spectacle before them. The film still provokes desire—the Major is a pornographic subject—but this type of desire does not fit nicely into psychoanalytic models premised on castration anxiety. Except for Togusa and Chief

---

103. Doane (1987) argues that women—as a gender—undermine masterful embodied viewing because their spectoricity frustrates narrative; their relation to language and the phallus is difficult, and their relation to desire mediated at best.

104. See Laura Mulvey, "Visual Pleasure and Narrative Cinema" in *Visual and Other Pleasures* (Bloomington: Indiana University Press, 1989), 14–26.

105. Doane (1987) claims that women's films also have an affinity with paranoid films, since both femininity and paranoia carry with them the sense of being on display.

Aramaki, it's not clear that *any* character has a penis—that inspiration for the phallus. This complication of desire—this denial of distance—means that perhaps, in this anime and others that feature female protagonists, the viewers and the female protagonists passively desire; they desire to desire. We are all therefore disembodied, and hence identify with computerization.[106]

This anime cyborg-subjectivity, which embraces castration and specularity *through* the representation of female bodies, coincides with Kaja Silverman's discussion of male subjectivity at the margins. Stressing the difference between the ego and the moi, Silverman argues that "moi" is a "psychical 'precipitate' of external images, ranging from the subject's mirror image to parental images to textually based representations we imbibe daily … what the subject takes to be its 'self' both other and fictive." Thus, notes Silverman, desire and identification are closely knit, since it is one's own ego that one loves in love.[107] Silverman also distinguishes between the gaze and the look. The gaze is always from outside—the subject never simply possesses the gaze; the gaze is that look from outside that constitutes the subject. One performs before the gaze, and in classical cinema, the male look is made to coincide with the gaze. We are, however, not simply the object of the gaze but rather also possessors of a look. We are therefore always both subject and object.

The simulated "camera work" highlights this difference between the gaze and the look as well as the connection between spectator and protagonist (this simulated camera work is often cited as the difference between anime and mainstream U.S. animation). When asked why he uses the "fish-eye" effects in anime, Oshii replied: "If you pressed me, you could say that these are the 'eyes' that look at the world of the film from the outside—that these are the eyes, in fact, of the audience."[108] Oshii, trying

---

106. The sound track—which is relentlessly Asian and female—furthers this female disembodying effect. I owe this insight to Jeffrey Tucker.

107. Kaja Silverman, *Male Subjectivity at the Margins* (New York: Routledge, 1992), 3, 4.

108. Carl Gustav Horn, "Interview with Mamoru Oshii," in *Anime Interviews*, 139.

to represent a networked society, links every scene and camera angle through a look or a sound, so that like a game of tag, someone is always "it"—someone's look coincides with the viewer's, more often than not the Major's. This constant move highlights the importance of the gaze and the look to fantasy. As Silverman remarks, "Fantasy is less about the visualization and imaginary appropriation of the other than about the articulation of a subjective locus—that is 'not an *object* that the subject imagines and aims at but rather a *sequence* in which the subject has his own part to play.'"[109] The last scene, in which the camera comes online again after fading out with the Major, makes explicit the viewer's role. When it does so, the audience jacks in as an audience with a line of vision that for the first time, does not coincide with anyone else's (only the "camera's"). This new line of vision brings to the fore the ambiguity of the frequent over the shoulder shots, where it is unclear whether a character is looking at the Major or the audience. In figures 4.30 and 4.31 (the dialogue between the Major and the Puppet Master), for instance, the Puppet Master could be addressing the audience when s/he says, "Because in you I see myself." In this manner, the viewer is another cyborg, a ghost who haunts the screen.

At the same time, this jacking in has a precedent within cyberpunk fiction itself—namely, Case's relationship with Molly. Case literally jacks into Molly, seeing what she sees and physically, if not emotionally, feeling what she feels. Similarly, the viewer jacks into the Major, and the portrayal of the Major's female connectors makes this explicit: in *Ghost in the Shell*, jacking into cyberspace is not portrayed as ejaculating into the system or penetrating the Net. Rather, the trodes emerge and penetrate the Major (see figures 4.32 and 4.33)—the cyborg is our female plug. The camera imitates the network connection, and when we look over the Major's shoulder, we take the position of a Net/console cowboy logging into her and seeing what she sees. This jacking in functionally parallels "passing" on the Internet. Rather than offering people an opportunity for others to lose their body or to "be" whoever or whatever they want to be, cyberspace as popularly conceived offers simstim—the illusion of jacking into

---

109. Silverman, *Male Subjectivity*, 6.

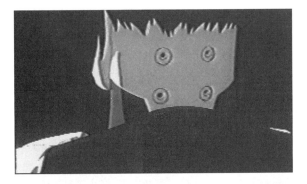

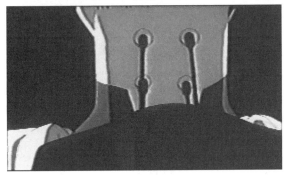

| Figure 4.32 |
The "female connectors" at the back of the Major's head

| Figure 4.33 |
Cyberspace jacked into the Major

another being, seeing what they see, and pretending to be who they are. There is always an option of jacking out, of leaving when things get too uncomfortable or difficult.

In order to effect such an insertion, anime viewers turn to cyberpunk fantasies about the Orient already in place, invariably a prerequisite to anime fandom. If "through fantasy, *'we learn how to desire,'*" through cyberpunk fantasies such as *Blade Runner* and *Neuromancer*, the viewer learns to desire and enjoy anime. The viewer identifies with protagonists such as Case and Deckard, who are faced with a world dominated by technology and all things Asian. The uncompromising nature of anime, the sense of being thrown into another culture and not being able to completely understand the situation, reiterates Case's position in *Neuro-mancer*. The arbitrariness of the trouble one finds oneself in, combined with the green cyberspatial views that makes everything comprehensible in terms of a cat-and-mouse chase, is exactly what anime offers its U.S. otaku viewers. The inability to comprehend Japanese and read all the signs afforded one, rather than alienating the viewer, places them in a position structurally mimicking cyberpunk heroes.

Does nationality affect this jacking-in effect? Newitz asserts that watching anime feminizes U.S. boys and thus places them in a capitulatory position to Japanese culture. They submit to that which they view and are overtaken by another's culture. Such a view assumes that feminization equals submission, ignores the fact that the viewer jacks in, rather than gets jacked into, but also, crucially, ignores the fact that anime's gaze feminizes its audience regardless of nationality. Importantly, otaku on both sides of the Pacific are considered effeminate or irregularly male.[110]

---

110. The socioeconomic status, age, and gender of mecha fans are similar on either side of the Pacific. As director Mamoru Oshii rather facetiously noted at an anime conference, there seemed to be little difference between his Japanese and U.S. audiences: "Both groups show a notable lack of females, and both seem to be the 'logic-oriented' type" (Carl Gustav Horn, "Interview with Mamoru Oshii," in *Anime Interviews*, 139). Specifically, both Japanese and U.S. audience members identify/are identified as otaku, which in Japan has become a derogatory term. Akio Nakamura first used the term otaku to describe attendees at a Komiketto (comic) convention. He writes that they "all seemed so odd ... the sort in every school class; the ones hopeless at sports, who hole up in the classroom during break ... either so scrawny they look like they're malnourished or like giggling fat white pigs with silver framed glasses with sides jammed into their heads ... the friendless type" (quoted in Schodt, *Dreamland Japan*, 44). In the United States, however, otaku has come to signify insider nerd-cool: as mentioned previously, *Wired* magazine's first cover featured the word otaku written in Japanese with no English translation, serendipitously placed next to a picture of Bruce Sterling's head. In fact, U.S. marketing strategies conflate anime with edgy cool. Eleftheria Parpis, in *Ad Week*, declares that "Japanimation is edgy and cool—and shops love it" ("Anime Action," 18). Analyzing Blockbuster's use of anime in its 1998 Christmas advertising campaign, she argues that "the ad targets the video game-playing-cartoon-watching 18–34-year old set; for them, Japanese animation is shorthand for insider cool" (20). Indeed, although anime in Japan stretch from historical drama feature films to children's television series, anime popular in the

Newitz's view would imply that U.S. otaku are especially feminized, but Newitz also suggests that translating and viewing anime may be a means by which viewers "convert a Japanese product into a uniquely American one. What might be satisfying for Americans about this is that it essentially allows them to 'steal' Japanese culture away from Japan."[111] This view supports the notion of anime as producing a Peeping Tom or spying effect. Indeed, Antonia Levi initially claims in *Samurai from Outer Space* that anime enables a great cultural exchange.[112] But Levi also argues that anime enables a penetrating view into Japanese society:

Anime can show you a side of Japan few outsiders ever even know exists. Unlike much of Japanese literature and movies, *anime* is assumed to be for local consumption only. That's important, because most Japanese are highly sensitive to outside pressure.... They write for and about Japanese. As a result, their work offers a unique perspective, a peeping Tom glimpse into the Japanese psyche.... But be warned. What you learn about Japan through *anime* can be deceptive. This is not the way Japanese really live. This is the way

U.S. market "generally fall into two broad categories: children's films and science-fiction adventures" (Charles Solomon, "For Kids, a 'Magical' Sampling of Japanese Animated Stories; Movies: UCLA Archive Caters to Growing Interest in Anime with Screenings of Features and Shorts," *Los Angeles Times*, January 8, 1999, 10). According to Matt Nigro of Manga Entertainment, "Most of our movies take place in the 21st century, follow futuristic sci-fi story lines and are geared towards 17- to 28-year-old males and females whose interests include music, comics, virtual reality, Internet surfing and computer games" (quoted in Rob Allstetter, "Entertainment: Japanese Videos Get Animated Interest," *Detroit News*, January 17, 1996, J1).

111. Newitz, "Magical Girls," 3.

112. According to Antonia Levi,

The new generations of both Japan and America are sharing their youth, and in the long run, their future. However much their governments may argue about trade and security in the Pacific, American's Generation X and Japan's *shin jinrui* will never again be complete strangers to one another. The connection is not only with Japan. *Anime* has already spread across most of Asia. Future social historians may well conclude that the creation of the American *otaku* was the most significant event of the post–Cold War period. (*Samurai from Outer Space: Understanding Japanese Animation* [Chicago: Open Court, 1996], 1–2)

they fantasize about living. These are their modern folk tales, their myths, their fables. This is not a peep into the conscious Japanese mind, but into the unconscious.[113]

The viewer, looking over the Major's shoulder, peeps into the Japanese unconscious, penetrating to the very ghost in the shell.[114] Anime as a great mirror, or illusion, enables one to look through a mirror darkly.

Thus, as Susan Pointon notes, "What is perhaps most striking about *anime*, compared to other imported media that have been modified for the American market, is the lack of compromise in making these narratives palatable."[115] Although the television series *Sailor Moon* was revamped for an English audience by changing the main character's name, Usagi (bunny), to Serena so that it would not offend female viewers, these anime do not go through an intensive Americanization before they hit the market. Indeed, among hard-core fans, the less mediated the better, and subtitled versions are valued over dubbed ones. This fetishizing of the other and the emphasis on incomprehensibility has not been lost on anime and manga creators. Rumiko Takahashi, the creator of *Ranma 1/2*, speculates that the popularity of anime in the United States may stem from exoticism: "Because I consciously feature Japanese life such as festivals and the traditional New Year's holiday, rather often in my manga, I sometimes wonder if American readers understand what they're reading. Maybe they just like the comics because they're exotic."[116] Exoticism and authen-

---

113. Ibid., 16.

114. As Frederik Schodt argues,

Ultimately, the popularity of both anime and manga outside of Japan is emblematic of something much larger—perhaps a postwar "mind-meld" among the peoples of industrialized nations, who all inhabit a similar (but steadily shrinking) physical world of cars, computers, buildings, and other manmade objects and systems. Patterns of thinking are still different among cultures, and different enough for people to be fascinated by each other, but the areas of commonality have increased to the point where it is easier than ever before to reach out and understand each other on the deepest levels of human experience and emotion. (*Dreamland Japan*, 339)

115. Susan Pointon, "Transcultural Orgasm as Apocalypse: *Urotsukidoji: The Legend of the Overfiend*," *Wide Angle* 19, no. 3 (July 1997): 35.

116. Horibuchi, "Interview with Rumiko Takahashi," 18.

ticity do appeal to viewers, and more often than not authenticity is proven by incomprehensibility. The *true* Japanese anime, unlike *Power Rangers*, do not try to address a non-Japanese audience. This insistence on anime as quintessentially Japanese and difficult to understand, as Thomas Lamarre has argued, constructs an essential Japaneseness that is untenable given the anime's position within global culture, given its own translation of animation.[117] It also assumes that U.S. culture is entirely readable.

Translation, however frugal, does appropriate another culture even as it establishes a bridge between cultures. In a translation, materials are domesticated—at the very least they must be rendered in one's domestic language and the domestic subject inserted. Yet, as Rey Chow drawing on Walter Benjamin has contended, translation (between media and languages) is a process of putting together and inscription that exposes the "original's" construction in all its violence; translation, which is not a one-way movement from an "original" to a "translation," is "a *liberation*, in a second language, of the 'intention' of standing-for-something-else that is already put together but imprisoned—'symbolized'—in the original."[118] This reciprocal liberation makes "both the original and the translation recognizable as fragments of a greater language, just as fragments are part of a vessel."[119] Chow, emphasizing the corruption of "original" and "translation," argues that a text's transmissibility depends on the level of its contamination: transmissibility "*intensifies* in direct proportion to the sickness, the weakening of tradition." Specifically analyzing translation between literary and visual texts, Chow maintains that the literalness of visual texts depends on their transparency, on what is "capable of offering itself to a popular or naive *handling*": "In the language visuality, what is 'literal' is what acquires a light *in addition to* the original that is its content; it is this light, this transparency, that allows the original/content to be transmitted and translated." The displacement of literary signification leads to a new way of thinking about one's native texts, "*as if it were a foreign culture peopled with unfamiliar others.*" Further pushing Benjamin's

---

117. Lamarre, "From Animation to *Anime*."

118. Chow, *Primitive Passions*, 187.

119. Walter Benjamin, quoted in ibid., 188.

contention that a translation is transparent—that it is an arcade rather than a building, Chow also claims that "the light and transparency allowed by 'translation' are also the light and transparency of commodification."[120]

The mutual and ongoing translation between thoroughly contaminated Japanese and U.S. cyberpunk—neither of which stands as an original—reveals each other's construction and the construction of this standing in called cyberspace (as they also erase the importance of other nations). Read together, they critique each other's Orientalism and expose the violence enabling their construction; they disorient each other's Orientalism, even as they rely on it to orient their own narratives. Each confronts/treats its "native" text "as if it were a foreign culture peopled with unfamiliar others"—making its audience "see" itself anew, but also exposing the violence inherent to constructing the foreign in the texts from which it draws. This mutual contamination not only describes the transmission between these texts but also the content of their narratives: the contamination of culture by technology, humanity by machines. And in many ways, it is this translation between that makes this other translation comprehensible (while at the same time obfuscating it). This multiple translation reveals the ways in which technology does not stand outside culture; rather, technology and culture constantly displace each other in a structure in which they are always made to stand in for each other.

The translation between media is as significant as the translation between languages. The movement from text to anime, from film to anime (in the case of *Blade Runner*) reveals textual and filmic construction. Oshii's re-creation of the look, for instance, reveals the work behind realist films—a look he (using technology) fuses with technology in order to reveal the difference mechanization makes. The translation to anime also graphically exposes the violence of Orientalist display, of viewing oneself as a spectacle (as it also uses spectacle). Perhaps unintentionally, anime also exposes the limitations the very imaginings of cyberspace—the ways in which cyberspace, that limitless land of possibility, has been constrained to repeat conventions. Crucially, Gibson's text is also a translation—a translation into text of video games, military technology, and popular

120. Chow, *Primitive Passions*, 199, 200, 19, 201.

visual culture. This translation—which is structurally a looking forward enabled by a looking backward—renders everything into a surface, a spectacle, and relies on the shiny light of commodities. Gibson's very transparency disorients and confuses the reader who must struggle with this shiny object that has none of the depth usually afforded by prose (and thus perhaps the saving grace of bodiless exultation). All this transparency, in other words, others. It creates what Chow drawing on Gianni Vattimo, drawing on Friedrich Nietzsche, calls a fabling of the world:

Instead of moving towards self-transparency, the society of the human sciences and generalized communication has moved towards what could, in general, be called the "fabling of the world." The images of the world we receive from the media and the human sciences, albeit at different levels, are not simply different interpretations of a "reality" that is "given" regardless, but rather constitute the very objectivity of the world. "There are no facts, only interpretations," in the words of Nietzsche, who also wrote that "the true world has in the end become a fable."[121]

This new fabling of the world is, as Nietzsche argued for the old, based on language, but also differs from the old because of transformations to language. It is not only the translation from language to image but also the translation from language into instrumental language (which increasingly produces these images, with whose surfaces we grapple). This other, invisible translation between voltages and signifiers, code and interfaces—obfuscated by visual culture—is also obfuscated by cyberpunk's mutual translation.

Going Native

The Orientalizing of the digital landscape, the entry into cyberspace as an entry into the world of Oriental sexuality, is not limited to literary and animated conceptions of the Internet, although Gibson's Orientalism, combined with Ridley Scott's, has become an enduring legacy in cyberpunk fiction: from Stephenson's *Snow Crash* and *The Diamond Age*, to the

---

121. Gianni Vattimo, quoted in Chow, *Primitive Passions*, 198.

Wachowski Brothers' *The Matrix*, every cyberpunk fiction contains some residual Asianness, even if its vision differs from *Neuromancer* or *Blade Runner*.[122] As discussed in chapter 2, Marty Rimm, whose senior thesis became the notorious Carnegie Mellon report on the consumption of pornography on the information superhighway, asserts that cyberspace introduces nine new categories of pornography, two of which are Asian and interracial. In this supposedly identity-free public sphere, not only has Asian pornography emerged as a popular genre but Asian itself has become a pornographic category.

The Internet also revises our understandings of Orientalism by disengaging Orientalism from the Orient. Through high technology, Orientalism is made to travel. Oriental mail-order bride sites such as Asian Rose Tours feature women from the former Soviet Union as well as the Philippines; when asian69.com first went online in 1999, it offered pictures of bound or mutilated white women. The conceit behind these sites is that Oriental women are submissive, and in some way lacking the independence and status of their white counterparts (the visitors to these sites are American *and* Japanese, among many other nationalities). The inclusion of Russian women exposes the economic base behind this assumption and the flexibility of the category Oriental. In 2004, mail-order bride sites were predominantly Eastern European. High-tech Orientalism, then, disperses Orientalism, in all the meanings of the word disperse. High-tech Orientalism seems to be all about dispersal, specifically the dispersal of global capitalism and networks.

These attempts to contain the Internet, to restrict it via Orientalism, do not guarantee safety. Orientalist narratives are not always comforting; they do not always orient. Rather, they carry with them fear of the yellow peril, or uncontrollable and contagious intercourse; they carry fears of overwhelming contact, of being taken over by the very thing they seek to

122. For more on the relationship between *The Matrix* and Japaneseness, see chapter 3 of Nakamura, *Cybertypes*. For more on the relationship between *The Matrix* and *Ghost in the Shell*, see Livia Monnet, "Towards the Feminine Sublime, or the Story of 'a Twinkling Monad, Shape-Shifting across Dimension': Intermediality, Fantasy, and Special Effects in Cyberpunk Film and Animation," *Japan Forum* 14, no. 2 (2002): 225–268.

control. They carry with them the fear of going native. As discussed in chapter 1, Senator Exon portrayed cyberspace as spreading obscene pornography, even though Exon, when arguing on the Senate floor for Internet regulation, had never surfed the Web for porn. Instead, he had a "friend" print off the most vile online pornography and then he carefully compiled it into a little blue binder, which he brought to the Senate chamber. Before the vote on the CDA, his peers came over to his desk, looked at the pictures, and then overwhelmingly supported the CDA. His notebook, in many ways, served as a perverse version of "look at my pictures from my friend's last vacation." Exon's horror at "hard-core" pornography and his desire to censor such materials parallels European reactions to excessive Oriental intercourse. As Said argues, "Every European traveler or resident in the Orient has had to protect himself from its unsettling influences.... In most cases, the Orient seemed to have offended sexual propriety; everything about the Orient—or at least Lane's Orient-in-Egypt—exuded dangerous sex, threatened hygiene and domestic seemliness with an excessive 'freedom of intercourse,' as Lane put it more irrepressibly than usual."[123] Again, what ruffles legislators' feathers about the Internet is freedom of intercourse in all senses of the word intercourse and in the dangerous sense of freedom. Faced with the information superhighway and the massive deregulation of the telecommunications industry in 1996, the government seized on pornography—excessive sexuality—as the reason for regulation.

But what happens when we take freedom of intercourse seriously, even if it is within the rubric of high-tech Orientalism? Consider, for instance, virtual sex. In many ways virtual sex epitomizes the Orientalist dreams of the Internet. As Cleo Odzer observes in *Virtual Spaces: Sex and the Cyber Citizen*, "Western men play with Thai prostitutes with the same non-chalance we play with our cyber-lovers."[124] The guiding metaphor of the Web—namely, virtual travel—feeds into the notion of the Internet as a vacation space, in which responsibility is temporarily suspended in favor

---

123. Said, *Orientalism*, 166–167.

124. Cleo Odzer, *Virtual Spaces: Sex and the Cyber Citizen* (New York: Berkley Books, 1997), 239.

of self-indulgence. Virtual sex seems always to verge on the "deviant": bondage, domination, sadism, and masochism dominate virtual sex, which furthers the theme of submissive and deviant Oriental sexuality.

Virtual sex and all so-called real-time communications cannot be safely cordoned off because they are not limited to the self, and because cyberspace cannot be limited to narratives of it perpetuated by works in other media that try to tell the truth about it. Instead, these real-time communications enable a form of contact that disables the notion of disembodied communication. The mirror starts breaking down. By now, we've all heard stories of people addicted to chat rooms and virtual sex— people whose lives and marriages have been destroyed by virtual infidelities or obsessions, or people whose definition of community has been redefined by their online participation. Further, rather than marking a disembodied space, the Internet creates spaces in which people pass, rather than imagine themselves as everywhere yet nowhere. In real time, dreams of exploration and domination are put to the test. The fact that real-time communications are never really real time, that there is a considerable time lag between question and response, also makes this space disorienting, and it is this disorientation, I argue, that enables the Internet to verge toward the disruptive, to verge toward the truly public. In real-time communications, narratives do not prevent contact with the "new."

Again, the Internet is not *inherently* Oriental but has been made Oriental, and high-tech Orientalism does not seal fiber-optic networks. The narrative of the Internet as Orientalist space accompanies narratives of the Internet as disembodied space. In other words, the Internet can only be portrayed as a space of the mind if there is an accompanying Orientalizing of difference, if there is an accompanying display of Orientalized bodies. Yet this binary of disembodied mind, on the one hand, and disembodied Orientalized other, on the other, breaks down with so-called real-time communication. This binary begins to break down in much cyberpunk fiction after Gibson, even if, influenced by Gibson and Scott, almost all cyberpunk to some extent uses Japaneseness to signal the future. Stephenson's half Korean, half African American protagonist—although a "case" —differs significantly from Case, and in his vision of the Metaverse, racial differences and representations of bodies proliferate. Although Cadigan's protagonists eat fast-food sushi, her vision of the future is not pinned to all things Asian; her fictions do not feature disembodied cowboy heroes either: in *Synners*, jacking into the Net (or more properly, being jacked by the Net) does not result in bodiless exultation. Rather, one can be jacked in and still grounded in one's body. Also, although not considered cyberpunk, Octavia Butler's *Patternmaster* series portrays mind-to-mind communications as disruptive and controlling: her empaths regularly commit suicide in order to escape. As well, in her dystopian *Parable* series, Butler presents "cyberspace" technologies as middle-class toys rather than tools "detourned" by the oppressed.[125] Importantly, this binary breaks down not because the Orientalized other is suddenly afforded the status as subject but rather because the boundary between self and other, self and self, freedom and control, begins to collapse.

---

125. See Dery, "Black to the Future."

| 5 |

CONTROL AND FREEDOM

Rosalyn Deutsche ends *Evictions: Art and Spatial Politics* by diagnosing a new form of *agoraphobia*. Manifested by those who long for consensus and rationality, this latest version masks fear as nostalgia: rather than admitting their fear of open spaces, these agoraphobics claim that the masses or identity politics have made post-(variable date) public spaces unlivable. This agoraphobic alibi thus "makes its narrator appear to be someone who is comfortable in, even devoted to, public space ... [while] it also transforms public space [from a space of contestation] into a safe zone."[1] It bypasses the question of whether or not such safe public zones ever existed by simply asserting that they have been "lost." Drawing on Thomas Keenan's contention that the public is always "elsewhere," Deutsch argues that these agoraphobics can never find these public zones because the public is phantom. Rather than existing as a space that one can inhabit, "it is the experience, if we can call it that, of the interruption or the intrusion of all that is radically irreducible to the order of the individual human subject, the unavoidable entrance of alterity into the everyday life of the 'one' who would be human."[2] Therefore according to Deutsch, the agoraphobics' nostalgia is a "panicked reaction to the openness and indeterminacy of the democratic public as a phantom."[3]

---

1. Rosalyn Deutsch, *Evictions: Art and Spatial Politics* (Cambridge: MIT Press, 1996), 326.

2. Thomas Keenan, "Windows: Of Vulnerability," in *The Phantom Public Sphere*, ed. Bruce Robins (Minneapolis: University of Minnesota Press, 1997), 133.

3. Deutsch, *Evictions*, 325.

Thus far, I have been arguing that different, and oftentimes conflict-ing, agoraphobic cover stories—which conflate freedom with control—underpin representations of fiber-optic networks as a public. These narra-tives vary from procensorship arguments that divide online contact into the good and the bad, to jingles that declare cyberspace a race-, gender-, age-, and infirmity-free marketplace of ideas, to Orientalist notions of cyberspace. The first holds that disruptive and invasive interactions online stem from certain types of content rather than from the Internet's open architecture. Accordingly, the Internet would be a safe space if only por-nography and other objectionable materials were eradicated. The second asserts that because electronic representations are nonindexical, they allow us to put the phantom public sphere to rest. By positing racism as "natu-rally" arising from physical differences, this assertion condemns physical space as irrevocably inequitable and refuses to register the ways that differ-ence affects one's representation (in both senses of the verb to represent).[4] All these narratives assume that private individuals precede public spaces, so that vulnerabilities result from contact with corrosive public air. As I have argued throughout, however, fiber-optic networks expose and in-volve us with others—human or otherwise—before we emerge as users. To follow Keenan's analysis, publicity functions as a language: language, he remarks, "intervenes on the lives of those who seek to use it with a force and a violence that can only be compared to … light, to the tear of the blinding, inhuman, and uncontrollable light that comes through a window—something soft, that breaks." Hence, "if we make images and express ourselves, we do so at the risk of the selves we so desperately long to present and represent."[5]

Fiber-optic networks may literalize Keenan's metaphor, but they do not simply extend the architectural or televisual window; they do not simply operate as a language. Or to be more precise, if they extend the window or operate as a language, they also stretch them beyond recogni-

4.   For more on the relation of difference to representation, see Rey Chow, "Gender and Representation," in *Feminist Consequences*, ed. Elisabeth Bronfer and Misha Kavka (New York: Columbia University Press, 2000).

5.   Keenan, "Windows," 138, 136.

tion. Fiber-optic telecommunications cables, double-coated glass tubes stretched to tiny threads, do not allow for vision. Unlike medical fiber optics, they do not simply allow for a more penetrating view. The "picture" we see on our screen is generated: there is no guarantee that the image we receive is a pixel-by-pixel representation of a previously recorded original. Sent over the network, our supposed representations—which can now only be understood as involuntary and voluntary user "events"—are sent in ways that cannot be seen or heard. Because of this, analyses of digital media that concentrate on the *appearance* of user interfaces or on high-level software miss what is fundamentally different about so-called computer-mediated communications: the fact that they are arguably human-mediated communications. Electronic traces are iterable, but not necessarily readable. As Wolfgang Ernst points out, digital media enables computers to talk to one another—the fact that humans can sometimes understand this ongoing conversation is due to specific translation programs.[6] Our fastly receding traces, mixed with and indistinguishable from inhuman traces and noise, follow the beat of an inhumanly precise drummer and create an archive that defies our senses.

Moreover, all electronic interactions undermine the control of users by constantly sending involuntary "representations." The routine, necessary, and nonceasing transmission of packets, which are constantly opened, broadcast, redirected, and possibly misdirected, simultaneously ensures that cyberspace can never conform to a "safe" marketplace and establishes these nonspaces as public. Users are *created* by "using" in a similar manner to the way drug users are created by the drugs they (ab)use. Paranoid narratives of Big Brother's all-seeing and all-archiving eye are similarly agoraphobic. They too mark as ideal noninvasive, happy spaces and also separate good from bad contact—as though the Web would be a safe space if only certain tracking mechanisms such as cookies were eliminated. The info-paranoid respond to the current "public" infrastructure of the Internet by creating private (that is, secret) spaces or cloaks, within which they hope to be invisible. They react to the systematic,

---

6.   Wolfgang Ernst, "Dis/continuities: Does the *Archive* Become Metaphorical in Multi-media Space?" in *New Media, Old Media: A History and Theory Reader*, eds. Wendy Hui Kyong Chun and Thomas Keenan (New York: Routledge, 2005), 105–123.

intrusive, and nonvolitional exchange of information by treating it as deliberately malevolent. Thus, just as new definitions of property combat digitization's seeming threat to private ownership, new definitions of privacy as secrecy emerge in response to electronic publicity. Dreams of human agency and subjectivity have not simply disappeared, although they too have been altered. Indeed, the very notion of cyberspace is a literary attempt to make sense of—to narrativize, to visualize, and hence to know—this seemingly unwelcome new public.

This structure of using while being used, this new exposure, exemplifies control-freedom, but also opens the possibility of a freedom beyond control. In this chapter, I elaborate on control-freedom, linking it to the rise of a generalized paranoia by revisiting the commercials addressed in chapter 3: paranoia drives the desire for technological objects and for control-freedom as a means of prevention. This chapter also examines control-freedom as propaganda/an algorithm of power by analyzing face-recognition technology and Webcams. It ends by arguing, with Jean-Luc Nancy, that freedom is something that cannot be controlled, that cannot be reduced to the free movement of a commodity within a marketplace. To do so is to destroy the very freedom one claims to be protecting; but unlike Nancy, it also insists on the importance of historical specificity, and on the links between decolonization and the increasing metaphoric significance of slavery.

Paranoid Musings

The Internet, as argued in chapter 3, was sold as empowering through commercials of happy people of color; however, it was also sold through a general (antiracist) racist paranoia. MCI's "Anthem" depends on seeing and knowing immediately, "of course *these* people would be happy to be on the Internet"; the solution offered (blindness as an antidote to faulty visual knowledge) screens our constitutive vulnerability and invasive technology. A jealous and paranoid logic drives these advertisements: it's not just "get online because it's happy there" but "hurry up and get online because all these other people are already there." You should want it because all these other people already want or have it. You should want it because of a mass-produced delusional advertisement. As chapter 3 contends, the demographic represented in this commercial was not representative, and race was always present on the Internet.

These commercials exemplify and perpetuate Jacques Lacan's conception of paranoid knowledge: the object (the Internet) is of interest to us because it is the object of another's desire. This paranoid knowledge propels the never-ending desire for technology within most advertisements/news reports—for technology that the other has, or we think the other has, which we think the other may use on us, which gives him or her an advantage over us. According to Lacan, "All human knowledge stems from the dialectic of jealousy, which is a primordial manifestation of communication." This "dialectic of jealousy" distinguishes the human from the animal world, paradoxically by making human objects proliferate neutrally and indefinitely: "There is no instinctual coaptation of the subject, in the way that there is coaptation, housing, of one chemical valency by another."[7] This lack stems from the fact that the object of human interest is the object of the other's desire, because

the human ego is the other and because in the beginning the subject is closer to the form of the other than to the emergence of his own tendency. He is originally an inchoate collection of desires—there you have the true sense of the expression *fragmented body*—and the initial synthesis of the *ego* is essentially an *alter ego*, it is alienated. The desiring human subject is constructed around a center which is the other insofar as he gives the subject his unity, and the first encounter with the object is with the object as object of the other's desire. . . . This rivalrous and competitive ground for the foundation of the object is precisely what is overcome in speech insofar as this involves a third party.[8]

Paranoid knowledge, based on jealousy and rivalry, stems from the mirror stage. Initially, the child is unable to distinguish itself from its environment and is an "inchoate collection of desires." During the mirror stage, the infant first imagines itself as unified rather than fragmented by recognizing its mirror image. This recognition, however, is a misrecognition,

---

7.   Jaques Lacan, *The Psychoses, 1955–1956: The Seminar of Jacques Lacan, Book III*, ed. Jacques-Alain Miller, trans. Russel Grigg (New York: W. W. Norton, 1993), 39.

8.   Ibid.

| Figure 5.1 |
Two happy African boys who tell us, "Three million more in the next five years," and later ask us, "Are you ready?"

for the mirror image, which appears as a more complete body, is not the child. The child therefore simultaneously identifies with and is jealous of its mirror image. This jealous identification launches the unending chain of identifications to follow, and in this aggressive identification with the other, we displace our own frustration onto the other's seemingly unified image, which we want to fragment.

Cisco Systems's "Empowering the Internet Generation" television commercial series illustrates this paranoid (visual) knowledge succinctly. These commercials, comprising different "scenes" (each scene is recycled in the different commercials), feature people from Asia, Southeast Asia, the Middle East, Africa, and Australia who address the audience and offer projections and statistics. In the "Networking Academies" commercial, these youngsters move (run, ride horses, swing, ride a boat) when they appear not to see us, and we move virtually (via the panning camera) when they stand still and say in variously accented English (see figures 5.1–5.3):

| Figure 5.2 |
Chinese girl on a bus, who repeats the African boys statement, "Three million more in the next five years"

There are over 800,000 job openings.

There are over 800,000 job openings for Internet specialists right now.

Three million more in the next five years.

Three million more in the next five years.

Seven out of ten students on the Web.

Seven out of ten students on the Web say they are getting better grades.

By the time I'm eighteen.

By the time I'm eighteen, I would like to get a job using Internet skills.

Are you ready?

Are you ready?

I'm ready.

| Figure 5.3 |

Southeast Asian girl declaring that "by the time I'm eighteen, I would like to get a job using Internet skills"

Are you ready?

We're ready.

Virtually all Internet traffic travels over the systems of one company.

The same one sponsoring networking academies all over the world.

Cisco Systems.

Are you ready?

Are you?

Are you?

The repetitious form and content of this commercial establishes joyous yet rivalrous identifications between these interchangeable others, and between these others and the most interchangeable of all: the viewer. The students repeat all the words, except for those describing Cisco Systems or the Internet. The viewer both celebrates and envies the mobility of these others: even though the viewer observes them with her panning camera and even though they interact with "primitive" modes of transport (walking, horses, boats), the viewer is stuck in front of the television set, while they prance around and ask, "Are you ready?" This envy makes the Internet desirable, especially to those who have yet to experience its banality. Predictably, Cisco Systems stopped running these commercials in the United States after the dot-bombs of 2001, running them instead in Japan, where Internet usage was still low. These commercials lure people onto the Internet with the threat of being left behind—they do not reassure people that everything will be OK.

The events of September 11, 2001 have put this paranoid knowledge into relief, revealing the lie behind MCI's and Cisco Systems's commercials, behind the rhetoric of the Internet as a happy public sphere, in which people of color are grateful and content to be equal only online. Technological empowerment and the threat of being left behind are no longer benign. These events also reveal the uneasy technologically defined boundary between self and other that this propaganda maintained, even as it pretended to transcend it: the other, who always threatened to have technology, was never really supposed to have it, or to be able to wield it against us. As Arvind Rajagopal argues, technology marks the boundary between civilization and its others to the extent that George W. Bush called the airplane-flying terrorists "cave dwellers."[9]

The threat of the Internet registered most strongly among those pundits who had condemned the CDA as unwarranted censorship, and who had taken comfort in liberal ideals of open discourse and access. Terrorism took over from pornography as the signal danger posed by the Internet, and the emphasis moved from bad *content* to bad *people*. Steven Levy's "Tech's Double-Edged Sword," which appeared in *Newsweek*, epitomizes

---

9. See Arvind Rajagopal, "Imperceptible Perceptions in Our Technological Modernity," in *New Media, Old Media*, eds. Wendy Hui Kyong Chun and Thomas Keenan, 277.

this switch. Describing the ways in which the victims on the planes and within the towers used cell phones to contact their loved ones, he notes:

THE RECIPIENTS of those calls ... are undoubtedly grateful for the final opportunity to hear those voices. But before we celebrate another irreplaceable use of wireless communications, consider this: according to government officials, within hours of the explosions, mobile phones of suspected terrorists linked to Osama bin Laden were buzzing with congratulations for the murderous acts. They use them, too.

The contrast dramatizes a long-recognized truism: modern technologies that add efficiency, power and wonder to our lives inevitably deliver the same benefits to evildoers. The Internet is no exception. On Sept. 11 the Net seemed like a godsend.... But there is also every likelihood that the terrorists had exploited the Internet as well, using easily available and virtually untraceable accounts on Yahoo or Hotmail, and meeting in ad hoc chat rooms.

Perhaps the terrorists cloaked their planning with cryptography, once an exotic technology, now a commonplace computer utility. Communications could also be shrouded with steganography (hiding messages between pixels of a graphic—a reputed bin Laden technique) or anonymizers (which make e-mail untraceable). Such tools are lionized by freedom-loving "cypher-punks," who have shrugged off potential dark-side usage as a reasonable trade-off for the protection that crypto can provide just plain citizens; as with cars and telephones, the benefits way overwhelm the abuses.[10]

In this reevaluation, Levy emphasizes the Internet's dark side as though suddenly discovering the downside of the Internet and the unpredictability of freedom. The cliché or truism that he uses to challenge the Internet's "exceptionality," rather than enabling a more critical or technically engaged critique, erases real-life experiences with it. The ringing cell phones of suspected terrorists were linked to their surveillance and "every likelihood" did not translate into evidence of steganography, but such facts did not impact these soulful reevaluations or the subsequent revoca-

---

10.   Steven Levy, "Tech's Double-Edged Sword," *Newsweek*, September 24, 2001, ⟨http://amsterdam.nettime.org/Lists-Archives/nettime-1-0110/msg00042.html⟩ (accessed May 1, 2004).

tion of civil liberties and public information, just as revenues did not affect stock prices in the late 1990s.[11] ICANN put security on its agenda; the U.S. government withdrew public information from its Web sites; the U.S. military demanded new security measures in order to prevent what it called an "electronic Pearl Harbor"; and the Senate passed sweeping new electronic surveillance measures.[12]

These so-called critical reassessments of the Internet, which reduce to "bad people do bad things with technology" and "good people do good things," obfuscate accidents and vulnerability; the one using the technology—or more properly, one's intentions—determines the result. Their emphasis on *who*—fundamentally the paranoid's question—also perpetuates paranoia to the extent that it constructs an other who always has technology, who always threatens, and no rational explanation can allay this fear. Paranoia cannot be dispelled by rational explanation because paranoia is all about rational explanation and meaning: for the paranoid subject, there is always meaning. The subject may not know the meaning, but the event is understandable in that it regards (pertains to and looks at) the subject.[13] Paranoia also drives these critical reassessments to the degree that they construct prevention as a technological, rather than a political, task (or to put it in more Lacanian terms, to the extent that they focus on meaning rather than truth). Technologically speaking, paranoia is a valid information-processing technique; not only are paranoid interpretations correct, but a paranoid's obsession with meaning, his or her pull to seemingly irrelevant terms, grounds prevention in the age of fiber optics. Automatic digital storage and networks enable a postevent

---

11. Although there is evidence that Osama bin Laden encrypted hand-delivered information saved on diskettes, there is no evidence that the terrorists used cryptography or steganography to transmit their plans over the Internet. They mainly seem to have used open chat rooms and unencrypted e-mail to exchange information.

12. The warnings of the dark side of technology and these surveillance measures were not new. The U.S. government had been loosely interpreting laws for years in order to trace e-mail and Web usage without a warrant, and the phrase electronic Pearl Harbor was first coined in 1993.

13. Lacan, *The Psychoses*, 75.

traceability that buttresses "prevention," for a digital mass of information can always be mined for warning signs read in, but not "read" (search terms only become self-evident after an event). Paranoia, inseparable from racial profiling, thus becomes a way of generating keywords in advance—a human response to an inhuman mass of information that belies rational analysis. As David Dill, the Stanford University computer science professor who launched a petition drive for a paper backup to electronic voting systems (without which recounts would be impossible), explains, "What people have learned repeatedly, the hard way, is that the prudent practice—if you want to escape with your data intact—is what other people would perceive as paranoia."[14] Or as Andy Grove, the CEO of Intel, wrote in his book of the same title "Only the paranoid survive." In the age of fiber optics, law enforcement officials will always appear negligent, unless they are inhumanly attentive, unless they react to every single piece of stored information.

The U.S. Department of Homeland Security's call to take everything seriously, to report all suspicious events and persons, essentially endorses and spreads paranoia, the cost and efficacy of which remains to be seen. Post–September 11, 2001, the authorities affirm and contact their citizens' excited nerves. And so, Senator Patrick Leahy (D-VT)—CDA opponent and anthrax target—stated on a public radio talk show on September 12, 2002, "I think we have to ask ourselves: Is it coincidence that we're seeing such an increase in West Nile virus or is that something that's being tested as a biological weapon against us?" The next day, the Florida inter-state was closed because a woman saw three Middle Eastern–looking men eating together and thought she heard them hatching a terrorist plot. Afterward, she told the press, "I hope I haven't caused someone prob-lems that really didn't do anything because I wouldn't want to cause someone problems. But at the same time I thought what if they really are doing something and I caught them?"[15] This generalized paranoia, which

14. Quoted in Dan Keating, "New Voting Systems Assailed," *Washington Post*, (March 28, 2003), A12.

15. Quoted in David Green and David Kidwell, "Federal Sources Say Terror-ism Threat by 3 Students Was a Hoax," *The Miami Herald*, Sept. 13, 2002, ⟨http://www.miami.com/mld/miami/4068519.htm⟩ (accessed October 1, 2003).

makes every citizen an eye and an ear for law enforcement, ignores the difference between possibility and probability, and almost no denial, especially by the accused, seems to shake paranoid convictions or rumors. The police, who found no explosives in the men's car, angrily stated, "If this was a hoax, they will be charged."[16] The men's apparent innocence and the woman's admission that they possibly "really didn't do anything" did not shake the police's conviction that these three medical students were in some way responsible.[17] Indeed, what evidence could sufficiently convince people that Al Qaeda did not use steganography? As one of the arrested men's parents told the press, "My son was born and raised here. I feel like we don't have freedom here anymore. Anybody can call anybody to make any kind of accusation. And the authorities treat you like you are a criminal."[18]

The Rodney King trial made stunningly clear this "preventive" tendency to see some people of color as criminals in advance. In the inverted view of the white paranoiac, as Judith Butler argues, the image of the unarmed King raising his hand before four armed white police officers was read not as an image of a man defending himself, nor of a man being jerked by tasers, but rather of a black male rising, as instigating violence, as asking to be beaten (see figure 5.4). This logic conflates prevention and

---

16. "Three Suspects Played Stupid Joke, Feds Say," *NBC6 South Florida*, ⟨http://www.nbc6.net/news/1667232/detail.html⟩ (accessed September 13, 2003).

17. According to their lawyer, "The men deny 'playing a trick' on Nurse Stone, who had reportedly been giving them suspicious looks. They declare that the phrase in question 'bringing it down' pertained to a car, owned by one of the men, which he wanted transported to Miami" ("'Bring It Down' Was about a Car, Students' Lawyer Says," CNN.com, ⟨http://www.cnn.com/2002/US/09/15/fla.terror.students⟩ [accessed September 13, 2003]). As Tim Moore, commissioner of the Florida Department of Law Enforcement, put it, "Not withstanding whether it was done in jest or if it was done on purpose, the result is the same … Floridians were in a state of alert and we spent hundreds of thousands of dollars" ("Three Men Freed after Being Held for Hours in Florida over 'Alarming' Comments," *Jefferson City News Tribune*, ⟨http://newstribune.com/stories/091402/wor _0914020027.asp⟩ [accessed September 13, 2003]).

18. Quoted in "Three Suspects."

| Figure 5.4 |
Still taken from the Rodney King video

retribution, for "according to this racist episteme, he is hit in exchange for the blows he never delivered, but which he is, by virtue of his blackness, always about to deliver," Butler observes, adding:

The attorneys proceeded through cultivating an identification with white paranoia in which a white community is always and only protected by the police, against a threat which Rodney King's body emblematizes, quite apart from any action it can be said to perform or appear ready to perform. This is an action that the black male body is always performing within the white racist imaginary, has always already performed prior to the emergence of any video. This identification with police paranoia culled, produced, and consolidated in that jury is one way of reconstituting a white racist imaginary that postures *as if* it were the unmarked frame of the visible field, laying claim to the authority of "direct perception."[19]

19.   Judith Butler, "Endangered/Endangering: Schematic Racism and White Paranoia," in *Reading Rodney King: Reading Urban Uprising*, ed. Robert Gooding-Williams (New York: Routledge, 1993), 19.

The white Simi Valley jury placed an imaginary frame around this image, in which King first aggressed and thus deserved such a savage beating. In an act of reading denying all reading, the jury identified with white paranoia and read King's actions as threatening the barrier enabling white autonomy—a barrier protected by the police.[20]

Lacan also links autonomy (or freedom) with paranoia. In an analysis bereft of references to institutional slavery or colonization, he claims that the discourse of complete autonomy perpetuates the master-slave relation or unfreedom in modern society. Although a certain mental breathing space seems indispensable to the modern man, this private space—which parallels John Stuart Mill's privacy bubble—is a delusion, which actually destroys freedom. This discourse, notes Lacan,

plays a part in the modern individual's presence in the world and in his relations with his counterparts. Surely, if I asked you to put this autonomy into words to calculate the exact share of indefeasible freedom in the current state of affairs, and even should you answer, *the rights of man*, or *the right to happiness*, or a thousand other things, we wouldn't get very far before realizing that for each of us this is an intimate, personal, discourse which is a long way from coinciding with the discourse of one's neighbor on any point whatsoever.[21]

According to Lacan, this discourse of freedom, precisely because it is vague, personal, and treasured, produces resignation and the abandonment of human rights. Because our discourses do not coincide, we, like Schreber, resign ourselves to reality, where reality in Lacan's words means the "renunciation of what is nevertheless an essential part of our internal discourse, namely that we have not only certain indefeasible rights but that these rights are founded on certain primary freedoms, which can be demanded for any human being in our culture," while still holding on to

---

20.   For more on this, see Grace Elizabeth Hale, *Making Whiteness: The Culture of Segregation in the South, 1890–1940* (New York: Pantheon Books, 1998). In it, she argues that segregation responded partly to white Southerners' shock at seeing black middle-class people who did not differ from them in clothing or speech.

21.   Lacan, *The Psychoses*, 133.

our personal discourse of freedom.[22] This personal discourse is the discourse of the alter ego, of that which fails to adjust to reality; it is "fundamentally biased and incomplete, inexpressible, fragmentary, differentiated, and profoundly delusional in nature." Importantly, Lacan begs the question of a nondelusional freedom (this question still remains and will be pursued later in this chapter) and provides no historical details. This elision of historical specificity enables him to valorize the symbolic as breaking paranoid knowledge and leading toward truth, effectively ignoring the relation between colonization and language elaborated by scholars such as Frantz Fanon. He also assumes that conflict automatically leads to delusion and abandonment, overlooking the "openness" endemic to all forms of signification.

The question of nonparanoid freedom aside, this personal discourse of freedom as complete autonomy drives the current conflation of freedom with security. The discourse of racial equality is also intertwined with this freedom.[23] For example, on September 26, 2001, the CEO of Visionics Corporation, Joseph Atick, argued on CBS's *Early Show* that face-recognition technology (FRT) helps prevent terrorist attacks, in part because "there is no chance for human error or 'racial profiling' because there is no need for a human operator to fixate on a particular person. The camera does it all automatically." FRT is an "enlightened alternative to racial and ethnic profiling."[24] As in the MCI advertisement, a limitation of technology—its inability to distinguish colors well, its sensitivity to lighting—is sold as an asset. Alan Dershowitz, Harvard University law professor and self-proclaimed civil libertarian, echoed Atick in a *New York Times* op-ed endorsement of national identification cards: "A na-

---

22.   Ibid.

23.   The wave of suspicion directed against South and Southeast Asians within the United States post–September 11, 2001 has tempered this "selling point." Still, racial profiling would fail to detect terrorists such as Richard Reid.

24.   Quoted in Jeffrey Rosen, "Being Watched: A Cautionary Tale for a New Age of Surveillance," *New York Times Magazine*, October 7, 2001, ⟨http://www.nytimes.com/2001/10/07/magazine/07SURVEILLANCE.html#⟩ (accessed October 10, 2001).

tional ID card could actually enhance civil liberties by reducing *the need for* racial and ethnic stereotyping. There would be no excuse for hassling someone merely because he belongs to a particular racial or ethnic group if he presented a card that matched his print and that permitted his name to be checked instantly against the kind of computerized criminal-history retrieval systems that are already in use" (emphasis added).[25] These technologies, Dershowitz argues, dispel the *need for* racial profiling by inhumanly screening everyone and comparing them to a set of "known" terrorists. The possibility that racial profiling is unnecessary and obfuscatory is not even considered. In the name of liberal racist antiracism, we are asked to accept more and more invasive—yet painless—technologies (see figure 5.5). This obviously posed image from Viisage's promotional materials implies that no one will be caught off guard or screened against one's will. All these arguments remarkably assume a technological system believed to be, and actually, fail-safe.

FRT supplements, and thus supplants, visual knowledge. Identix, in its promotional materials, alleges that FRT partakes in the "eternal" use of human faces to verify identity.[26] If technology, especially computer simulation, disconnected knowledge and vision by reproducing productions rather than reproductions, FRT promises to reconnect seeing, knowing, and believing.[27] FRT, however, corrects for visual subjective bias by inhumanly bypassing rationalization and deduction: a terrorist is a terrorist not because he looks like one, nor because he carries certain weapons, nor because of past crimes, but rather because s/he can be positively matched to an already existing picture. Most face-recognition

---

25. Alan M. Dershowitz, "Why Fear National ID Cards?" *New York Times*, October 13, 2001, ⟨http://www.nytimes.com/2001/10/13/opinion/13DERS.html⟩ (accessed October 14, 2001).

26. "Understanding Biometrics: Face Recognition Technology," Identix, ⟨http://www.identix.com/newsroom/news_biometrics_face.html⟩ (accessed September 13, 2003).

27. This knowledge was put into question, though much earlier, with the emergence of urban crowds. As I argued more fully in chapter 1, urban crowds in the nineteenth century led to a desire for a "navigability" based on "readable" codes such as race and facial features.

| Figure 5.5 |
Viisage face recognition promotional materials

technologies rely on the "eigenface" technique, which drastically simplifies the necessary computation. The system determines a series of eigenvectors (called eigenfaces) for any given library of normalized images: every face in the library can thus be represented as a weighted combination of these eigenfaces (see figure 5.6).[28] To "recognize" a face, the system determines its coefficients and compares it to its closest match in the already existing coefficient matrix; the face "matches" if the error is sufficiently low. Rejection and recognition rates are negatively correlated: the

28.   The eigenface technique relies on basic linear algebra—namely, the fact that a coefficient matrix (multiplied by a variable matrix) can be represented as a sum of eigenvectors, multiplied by scalars (eigenvalues).

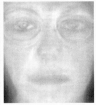

| Figure 5.6 |

Average image plus first four eigenfaces taken from Lemieux and Parizeau

more accurate the system, the more faces labeled unknown; the more faces recognized, the less accurate the system.[29]

Importantly, the primary developers of the eigenface technique see it as bringing "information theory" to the problem of face recognition.[30] Prior methods, relying on conventional human features, such as noses or eyes, or the distance between them, "ignored the issue of just what aspects of the face stimulus are important for identification, assuming that predefined measurements were relevant and sufficient."[31] These older humanly determined measurements stem from Alphonse Bertillon's nineteenth-century attempts to solve the "one-to-many" problem (specifically, catching repeat offenders) induced by photography. Bertillon created a filing system to archive mug shots (which he also standardized), based on eleven "sufficiently singular" measurements of the human body. These measurements created a grid, the smallest unit of which was twelve cards.[32] As Allan Sekula writes, "For Bertillon, the mastery of the criminal body necessitated a massive campaign of *inscription*, a transformation of the body's signs into a *text*, a text that pared verbal description down to a

29. Matthew A. Turk and Alex P. Pentland, "Face Recognition Using Eigenfaces." *Proceedings of the IEEE Conference on Computer Vision and Pattern Recognition*, Maui, Hawaii, 1991, 590.

30. Ibid., 586.

31. Ibid., 587. These features can be used to refine the method.

32. See Allan Sekula, "The Body and the Archive," *October* 39 (1986), 28.

denotative shorthand, which was then linked to a numerical series."[33] In contrast, FRT reduces the body immediately to a numerical series. The computer does not "see" these eigenfaces but rather reads and manipulates a series of vectors.

How well the computer "sees" is unclear. As many investigators admit, FRT more accurately recognizes (one-to-one) than identifies (one-to-many). FRT works best with face-on images, normalized for size and light, which airport surveillance cameras do not produce. The American Civil Liberties Union, in its report "Drawing a Blank," spotlights the fact that Tampa's police department, the first U.S. department to use FRT, has yet to make an arrest with it.[34] As well, camera location is crucial to FRT's racial "blindness." According to various reports on British surveillance camera use, surveillance cameras have worked mainly by displacing criminal activity and inducing paranoia. The camera's blindness also exceeds its framing, for the very act of observation, as Christian Katti has argued, creates a blind spot—a position from which it is impossible to see because no observer can observe "his own observing nor himself as observer."[35] Every observation produces its opposite: concealment.

FRT also fails because its libraries cannot be comprehensive. Terrorists are constantly recruited; one is not simply born a terrorist (this repeats Schreber's critique of the writing-down system—namely, that human thought cannot be exhausted and archived).[36] This fact also invali-

---

33.   Ibid., 33.

34.   ACLU, "Drawing a Blank," ⟨archive.aclu.org/issues/privacy/drawing_blank .pdf⟩ (accessed May 1, 2004).

35.   Christian Katti, "'Systematically' Observing Surveillance: Paradoxes of Observation according to Niklas Luhmann's Systems Theory," in *CTRL [SPACE]: Rhetorics of Surveillance from Bentham to Big Brother*, eds. Thomas Y. Levin et al. (Cambridge: MIT Press, 2002), 58.

36.   Such techniques, however, work quite well with legally endorsed "entrapment," and the current use of image-recognition technology to track pedophilia is an excellent example of this. In order to "identify" child pornography, the image is compared to a known set of child pornographic images—a set in all probability disseminated, if not produced, by law enforcement officials in their "proactive" approach to law enforcement.

dates the criticism leveled against FRT and other identification systems, from national identification cards to firearm registration, that they would only intrude on the privacy of law-abiding citizens, rather than register criminals (because criminals, who are born criminals, would naturally avoid all forms of documentation). By asserting, with Mill, that "good" people require private space, this argument does trouble the open/ good and closed/bad dualism, but also endorses it by assuming that criminals naturally seek the dark. A national identification system would register criminals, just as a firearm registration system would register guns used in violent crimes. The Federal Bureau of Investigation did have photographs of many of the 9/11 terrorists. Yet archiving (categorizing, recording, and searching in a systematic manner) and retrieving these images is still difficult, if not impossible. So for now, the best place to hide is within the mass of data compiled by information technologies —a mass that as Wolfgang Ernst notes, seems to be unarchivable. This failure, however, rather than pacifying paranoia, induces it. The inadequacies of control mechanisms, as Katti has similarly contended, induces paranoia.[37]

Paranoid Authority

Inadequate information, combined with an obsession with meaning, drives paranoid behavior.[38] Paranoia stems from the desire to compensate for a perceived weakness in symbolic authority. According to Eric Santner, Schreber's paranoia stemmed from his knowledge of the "rottenness" at the core of power—namely, that a performative imperative, which spawns

---

37. According to Christian Katti, the uncertainty generates a second-order observation that causes the paranoid subject to question the reality of first-order observation and reality itself ("Observing Surveillance," 58–59). Since it cannot observe itself observing, though, other distinctions cannot be seen without another extrapolation and such ordering can go on forever. This system would also not alleviate paranoia because catching would-be criminals/terrorists is not simply an issue of identification—they can be recruited at anytime.

38. Stanford University researcher Philip G. Zimbardo was able to produce paranoid behavior in subjects by inducing partial hearing loss and then placing them in a room full of people (Wray Hebert, "Paranoia: Fearful Delusions," *New York Times Magazine*, March 19, 1989, 62). I thank James Der Derian for this reference.

endlessly repeated rites of institution, lies at the heart of the legal system and every symbolic investiture. Schreber's paranoia, Santner argues, arising from a "crisis of investiture" (first, his failed election effort, and second, his promotion to presiding judge of the Saxon Supreme Court), bears testimony to the usually repressed relationship between the liberties and the disciplines.[39] Thus, paranoia does not respond to an overwhelming, all-seeing power but rather to a power found to be lacking—rotten and inadequate, always decaying. Paranoid knowledge similarly responds to technologies' vulnerabilities, even as it denies them. Paranoia increases as visibility decreases.

According to Lacan, a subject becomes paranoid when it cannot move from the imaginary into the symbolic, when it fails to undergo successfully the Oedipus complex: Lacan claims that Schreber was paranoid because he was missing a primordial signifier or "quilting point"—namely "being-a-father," a condition usually caused by unilateral and monstrous fathers. Usually, one can still operate without this primordial signifier, for "one missing a primordial signifier is like a three-legged chair; or if primordial signifiers act as highways—that is, direct paths from A to B—the absence of them makes one travel on smaller, less direct routes.[40] Such a person is left with the *image* of the paternal function "whose function as model, as specular alienation, nevertheless gives the subject a fastening point and enables him to apprehend himself on the imaginary plane." This image, however, is not "inscribed into any triangular dialectic ... [but rather] installed on a plane that has nothing typical about it and is dehumanizing because it doesn't leave any place for the relation of reciprocal exclusion that enables the ego's image to be founded on the orbit given by the model of the more complete other." This person, Lacan argues, will go

39.   Eric Sartner, *My Own Private Germany: Daniel Paul Schreber's Secret History of Modernity* (Princeton, NJ: Princeton University Press, 1996), xii.

40.   Lacan, *The Psychoses*, 203. According to Lacan, the most palpable example of a quilting point is the notion of father introduced through the Oedipus complex (268). There is no "natural" relationship between the signifier father and the signified, so this connection is made via the Oedipus complex. As I mention later, this does open the question: Now that DNA testing can determine this relationship nonsymbolically, is it still a quilting point?

through life compensating via "a series of purely conformist identifications with characters who will give him the feeling for what one has to do to be a man."[41]

The onset of psychosis and hallucinatory delusions, and thus the diagnosis of paranoia, happens when this lack

manifests itself through fringe phenomena in which the set of signifiers is brought into play. A great disturbance of the internal discourse, in the phenomenological sense of the term, comes about and the masked Other that is always in us appears lit up all of a sudden, revealing itself in its own function, for this function is the only one that henceforth maintains the subject at the level of discourse which threatens to fail him entirely and disappear.[42]

The onset of psychosis occurs when the subject encounters the signifier as such (in Schreber's case, when following an explicit call from the ministers to essentially be a father, to be the name of authority) and he cannot respond to its interpellation.[43] After such an encounter, the imaginary relation no longer suffices and the subject's entire signifying structure disintegrates. During this period, delusions occur, for that which has been foreclosed in the symbolic reemerges in the real.[44] After considerable effort, the signifying structure is reconstituted and the discordant signifier harmonized via extraordinary explanations.

Agreeing with this Lacanian framework, Slavoj Žižek contends that we are all paranoid. As paternal authority wanes, "the big Other" disintegrates:

The paradoxical result of the mutation in the nonexistence of the big Other—of the growing collapse of symbolic efficiency—is … the proliferation of

---

41.   Ibid., 205.

42.   Ibid.

43.   Ibid., 321.

44.   Lacan explains the symbolic through computers: just like computers, what goes into the symbolic must be formatted in a certain way and what is not formatted properly returns in the real.

different versions of a *big Other that actually exists, in the Real*, not merely as symbolic fiction. The belief in the big Other which exists in the Real is, of course, the most succinct definition of paranoia; for this reason, two features which characterize today's ideological stance—cynical distance and full reliance on paranoiac fantasy—are strictly codependent: the typical subject today is the one who, while displaying cynical distrust of any public ideology, indulges without restraint in paranoiac fantasies about conspiracies, threats, and excessive forms of enjoyment of the Other.[45]

According to Žižek, the growing collapse of symbolic efficiency—of enunciations that draw their hypnotic force from their very enunciation—makes us, like Schreber, paranoid.[46] In order to combat these proliferating real Others like Bill Gates, Žižek argues for "a return to the primacy of the economy"; that is, recalling the big Other to exorcise real ones.[47] Slightly revising Žižek's analysis, it could be argued that we are prepsychotic paranoids, caught in the plane of meaning and constantly seeking to compensate for the lack of authority through smaller *imaginary* ones. Or following Fredric Jameson's linking of schizophrenia and postmodernism, we are all schizophrenic, unable to quilt together signifier with signified, and thus caught in ever swirling affects—a condition, Chela Sandoval contends, normal for the oppressed. Perhaps.

But what is the difference between generalized paranoia and a clinical case, between theoretical and paranoid explanation? In common parlance, paranoia is not schizophrenia, for paranoid and "normal" explanations

---

45. Slavoj Žižek, *The Ticklish Subject: The Absent Centre of Political Ontology* (London: Verso, 1999), 362.

46. As Slavoj Žižek asserts, symbolic efficiency

concerns the minimum of "reification" on account of which it is enough for us, all concerned individuals, to know some fact to be operative—"it," the symbolic institution, must also know/"register" this fact if the performative consequences of stating it are to ensue. Ultimately, this "it," of course, can be embodied in the gaze of the absolute big Other, God Himself.... "Symbolic efficiency" thus concerns the point at which, when the Other of the symbolic institution confronts me with the choice of "Whom do you believe, my word or your eyes?" I choose the Other's word without hesitation dismissing the factual testimony of my eyes. (*Ticklish*, 326–327)

47. Ibid., 156.

differ only in degree: one injects "now, you're paranoid" when an explanation has gone too far. The paranoid's "mistake" is its belief in intention (Foucault claims that the paranoid errs by asking "who" rather than "what"); however, does the normalization of paranoia not indicate that the symbolic, privatized, and permutated through biology, computer technology, and seemingly unending copyright extensions, can no longer establish some third relation, and that sexuality, although still intertwined with language, is less discursive? In other words, does Žižek's belief that reasserting symbolic paternal authority will reinforce symbolic authority not obscure profound changes to the symbolic that cannot be rectified by patriarchy? (For instance, can "being a father" stand as a primordial signifier now that fatherhood can be scientifically determined?) And what is the relationship between paranoia, control technologies, and freedom? The question of language—and particularly the reduction of language to commands—would seem key.

William Burroughs, who inspired Gilles Deleuze's theorization of control societies, in many ways epitomizes the cynical distance and full reliance on paranoiac fantasy diagnosed by Žižek. Burroughs's notion of control also stems from control systems and addiction. In his 1961 interview with Gregory Corso and Allen Ginsberg, Burroughs, grandson and namesake of the inventor of the Burroughs Adding Machine, stated that "the machine should be eliminated. Now that it has served its purpose of alerting us to the dangers of machine control."[48] Although machines and scientists have perfected control, Burroughs maintains that control arises from words (which he viewed as viruses): "Suggestions are words. Persuasions are words. Orders are words. No control machine so far devised can operate without words, and any control machine which attempts to do so relying entirely on external force or entirely on physical control of the mind will soon encounter the limits of control."[49] Words thus do not stem from the need to communicate but rather the need to

---

48.   Quoted in Allen Hibbard, ed., *Conversations with William S. Burroughs* (Jackson: University Press of Mississippi, 1999), 3.

49.   William Burroughs, *The Adding Machine: Selected Essays* (New York: Henry Holt, 1986), 116.

control animals capable of resistance.[50] For Burroughs, as for Norbert Wiener, all language is commands, which are necessary because of resistance or acquiescence. Without resistance, one is "used" rather than controlled.[51] This implies that control requires free will, and so what we take to be freedom, the ability to decide, is the basis for control (this can be seen most starkly in birth control as sexual freedom and as eugenicists' dream—the ways in which, as Gayatri Spivak argues, northern feminists' insistence on abortion rights belittles the agency of southern women and covers over the complex relationship between poverty and "fertility").[52] Indeed, Burroughs eventually viewed freedom as escaping addiction and past conditioning by controlling the scientific apparatuses of control (he moved from calling for the death of all scientists to arguing that all scientific discoveries should be made public). Burroughs's own addiction and his apomorphine "cure" made him the perfect seer of an age in which one's own body would become the source of oppression.

The noninnocence of language, however, has been most forcefully explained by Frantz Fanon. Language, as Fanon repeatedly points out, is not innocent; it is not, as Lacan asserts, simply the introduction of a third term that breaks paranoid knowledge and starts us toward truth (or to respond to Thomas Keenan, language does not expose equally).[53] According to Fanon, speaking means taking on "the weight of a civilization," a po-

---

50.   Hibbard, *Conversations*, 195.

51.   This distinction would become less clear in his later writings. For more on this, see Burroughs's 1971 interview with *Penthouse*. (Hibbard, *Conversations* 39–50).

52.   See Gayatri Spivak's, "Public hearing on crimes against women," *Women Against Fundamentalism* 7 (1995), ⟨http://waf.gn.apc.org/journal7p3.htm⟩ (accessed September 3, 2004). The relationship between control and eugenics stems from the movements for "social control" that led to modern sociology.

53.   Frantz Fanon argues, "Like it or not, the Oedipus complex is far from coming into being among Negroes" because "every neurosis, every abnormal manifestation, every effective erethism in an Antillean is the product of his cultural situation," whereas Oedipal-inspired neuroses stem from the family. Fanon could also make this argument using language and Lacan (*Black Skin, White Masks*, trans. Charles Lan Markmann [New York: Grove Press, 1967], 152). Lacan states that

tentially crushing load for the colonized, who are "offered" an inferior "symbolic position":

Every colonized people—in other words, every people in whose soul an inferiority complex has been created by the death and burial of its local cultural originality—finds itself face to face with the language of the civilizing nation: that is, with the culture of the mother country. The colonized is elevated above his jungle status in proportion to his adoption of the mother country's cultural standards. He becomes whiter as he renounces his blackness, his jungle.... The black man who has lived in France for a length of time returns radically changed. To express it in genetic terms, his phenotype undergoes a definitive, an absolute mutation.[54]

---

if psychoanalysis teaches us anything, if psychoanalysis constitutes a novelty, it's precisely that the human being's development is in no way directly deducible from the construction of, from the interferences between, from the composition of, meanings, that is, instincts. The human world, the world that we know and live in, in the midst of which we orientate ourselves, and without which we are absolutely unable to orientate ourselves, doesn't only imply the existence of meanings, but the order of the signifier as well.

If the Oedipal complex isn't the introduction of the signifier then I ask to be shown any conception of it whatever. The level of its elaboration is so essential to sexual normalization uniquely because it introduces the functioning of the signifier as such into the conquest of the said man or woman. (*The Psychoses*, 189)

According to Lacan, the Oedipus complex introduces the signifier (specifically the name-of-the-father) in order to explain phenomena that may map onto the biological world, but that cannot be reduced to it. So the Oedipus complex necessarily introduces the name-of-the-father signifier because there is (or at least was) no way of verifying or explaining being-a-father except through symbolization. If one's entry into language enables "truth" and a position within the symbolic network, Fanon contends that not all positions are equal and that colonization uses this subjection to build complicity as well as inferiority. Fanon thus seeks to "help the black man to free himself of the arsenal of complexes that has been developed by the colonial environment" (*Black Skin*, 30). Gilles Deleuze and Felix Guattari also link the Oedipus complex to colonialism in *Anti-Oedipus: Capitalism and Schizophrenia*, trans. Robert Hurley, Mark Seen, and Helen R. Lane (Minneapolis: University of Minnesota Press, 1983).

54.   Fanon, *Black Skin*, 19.

Through language, the colonizer "fixes" the colonized as inferior; the colonizer speaks down to the colonized, assuming the colonized to have no culture or language (indigenous or otherwise), no matter how well he or she speaks. The colonized's "perfect" and defiant French makes the colonizer pause and sense that something is new under the sun; but the so-called face-to-face encounter (which liberal philosophy argues founds ethics) and the racial epithet fixes the colonized once more, replacing the black person's corporeal scheme with a racial-epidermal one.

Importantly, the current formulation of control as freedom claims to eradicate the open racism Fanon describes. In Burroughs's terms, it recognizes the formerly colonized as humans capable of resistance and acquiescence. Now, the call "Look, a Negro!" in a public French train, which Fanon argues replaces the black person's corporeal scheme with a racial-epidermal one, would not meet with the liberal tolerance Fanon depicts, and the continuing exploitation (which includes military and/or economic domination and occupation) in formerly colonized areas differs from colonialism. These "operation freedoms" pirate the freedom advocated by the decolonization and U.S. civil rights movements in order to implement control: freedom becomes what you cannot not want—not only because the desire for freedom is everywhere but also because those seeking to "free" do not allow anyone to want anything else. There is no other choice than their freedom, which is the freedom to be an individual—to exceed one's culture in order to become incorporated into a global market.

This freeing makes decolonization metaphoric and does not use overtly racist terms. Although racism always mixed together biology and culture, the post–World War II erosion of *overtly* racist biology—the loss of the racial epidermal schema as a valid form of visual knowledge—has changed the terms of discrimination. Race, in biological terms, is now what you cannot see, what cannot be expressed, yet it still persists as a way of trying to understand the invisible through the visible, falsely or not. Race persists as the frenzy of and decline in visual knowledge.

Freedom: What You Cannot Not Want

Freedom as something one cannot not want is key to control as freedom. Clearly, ideologies and practices of freedom and control are not new to this period, and control has been previously linked to liberty/freedom although as an opposite. To do something without control is to do it "freely"; John Stuart Mill viewed public opinion as a form of control operating against liberty (unless it punished an infringement against one's interest). Isaiah Berlin, distinguishing between negative freedom (the freedom to do what one wishes within a limited space—a freedom associated with the "free world" and the separation of public from private spaces) and positive freedom (associated with the former Communist bloc), did link self-mastery to positive freedom. Starting with Kant, Berlin argues that positive freedom splits the higher from the lower self: the higher self follows reason and the lower self the passions—in order to become a master, one must emancipate oneself from one's passions. This split opens up the possibility for abuse. Since Kant and others following him, such as Spinoza, Locke, Hegel, and Marx, believed that no rational law was against freedom, others could be subjugated in the name of freeing their "higher" selves. Positive freedom, Berlin claims, leads to totalitarianism because it assumes that the goals of humankind are one, but human ends, Berlin asserts, are diverse and rivalrous; humanity must always choose between eternal values, between negative freedom, equality, and fraternity.[55]

---

55. This separation of freedom from equality and fraternity is problematic. Arguing against the contemporary belief that all improvements to humanity's condition are a form of "liberation," Berlin claims that most decolonization efforts are not movements for freedom (although their desire to emerge as agents is akin to freedom) but rather demands for recognition, since these people prefer a dictator of their own race over a good administrator from a higher civilization:

When I demand to be liberated from, let us say, the status of political or social dependence, what I demand is an alteration of the attitude towards me of those whose opinions and behavior help to determine my own image of myself. What oppressed classes or nationalities as a rule demand is neither simply unhampered liberty of action for their members, nor, above everything, equality of social or economic opportunity, still less assignment of a place in a frictionless, organic state devised by the rational lawgiver. What they want, as often as not, is simply recognition (of their class or nation, or colour or race) as an independent source of human activity, as an entity with a will of its own, intending to act in accordance with it (whether it is good, or legitimate, or not), and not to be ruled, educated, guided, with however light a hand, as being not quite fully human, and therefore not quite fully free. (*Two Concepts of Liberty: An Inaugural Lecture Delivered before the University of Oxford on 31 October 1958* [Oxford: Clarendon Press, 1958], 41–42)

———

The current twinning of control and freedom reveals the lie behind Berlin's easy separation of negative from positive freedom, for negative freedom is now intertwined with control (although to even make the argument that negative freedom preserves choice, Berlin had to ignore the fact that Mill himself maintained that man cannot have the choice to refuse freedom).[56] George W. Bush, in his Address to the Nation in June 2002, stated, "Freedom and fear are at war. And freedom is winning.... Homeland security will control our borders and prevent terrorists and explosives from entering our country."[57] But freedom produced by homeland security is one based on fear: a gated community writ large. The hallmarks of this freedom are "securing cockpits, tightening our borders, stockpiling vaccines, [and] increasing security at water treatment and nuclear power plants" as well as media self-censorship and the creation of a national TIPS program, reminiscent of the Reign of Terror, from which the word terrorism stems.[58] The gathering and sharing of "intelligence"—the du-

---

This assessment of the struggle for decolonization and the spread of Marxism during the mid- to late 1950s is astonishingly naive, and takes as representative the position of native intellectuals. As well, by dividing freedom from its "sisters" fraternity and equality, Berlin is able to preserve freedom as a value belonging to the elite civilizations and also keep his dichotomy in place, since these movements fulfill all the requirements of positive freedom, but do not claim all in the name of reason.

56.   Berlin elides an important section of Mill's *On Liberty* (Chicago: Gateway Editions, n.d.). According to Berlin, choice rules negative freedom, for "to threaten a man with persecution unless he submits to a life in which he exercises no choices of his goals; to block before him every door but one, no matter how noble the prospect upon which it opens, or how benevolent the motives of those who arrange this, is to sin against the truth that he is a man, a being with a life of his own to live" (*Two Concepts of Liberty*, 12). Mill's own text, though, denies one fundamental choice: "The principle of freedom cannot require that he should be free not to be free. It is not freedom to be allowed to alienate his freedom" (*On Liberty*, 131).

57.   Quoted in "Fighting Terror: Government Overhaul President's Address; 'Freedom and Fear Are at War'" *Boston Globe*, June 7, 2002, 3rd ed., A37.

58.   George W. Bush, "Remarks by the President in Address to the Nation," June 6, 2002 ⟨http://www.dhs.gov/dhspublic/display?content=169⟩. For TIPS,

plication of the real world—is key to freedom as control, and this freedom, rather than securing the separation of public and private, eradicates
privacy for the average person: although the Bush administration operates
in a shroud of secrecy, it insists that those who are against its invasive programs must have something to hide.[59] The so-called free world now
espouses a version of freedom closer to positive than negative freedom:
the ends of humankind are the same (security)—the means are therefore
merely technological. In this unending war for freedom and against terror
(the Department of Homeland Security is a permanent department, and
has taken over the responsibilities of the Immigration and Naturalization
Services), everyone is a soldier for freedom, called on to sacrifice their
civil liberties and indulge in Las Vegas holidays or other freedom-loving
activities.

This paranoia thus also enables enjoyment, for us as for Schreber, in
Las Vegas or not; it creates a new intensified body through which we negotiate control and freedom. The most stark example of this are Webcams, which seemingly turn everyday people into celebrities by turning
surveillance into sexual and other pleasure. Indeed, many scholars view
Webcams and reality television shows as complementing surveillance
cameras.[60] According to Ursula Frohne, Webcams are a total surrender
of the individual to the gaze: "Observation and surveillance [have] become
the prerequisite for recognition" so that "whoever is not 'on the air' is
denied existence. We only experience ourselves as real when we are able
to make an 'appearance.' A sense of our own life becomes tangible only

---

individuals such as utility repair people are asked be the ears and the eyes of law
enforcement.

59.   One of the first tasks of the new Homeland Security Department was the
integration of various governmental databases, something that could have been accomplished much earlier, but was until recently decried as totalitarian.

60.   Significantly, most critics—Wolfgang Ernst being an important exception—
treat films that represent reality television (*The Truman Show*), television shows
(*Big Brother*), and Webcams indifferently, thereby erasing the difference between representation and object—a difference they claim the media they analyze
erases.

when it is reproduced. The media have therefore become the last authority for self-perception, the 'reality test' of the social persona: I am seen, therefore I am."[61] Peter Weibel observes that the "I am seen, therefore I am" moves exhibitionism and voyeurism from personal pathologies to everyday social conditions.[62] This diffusion of exhibitionism and voyeurism correlates with a neutralization of surveillance; Wolfgang Ernst argues, "No longer is panoptic surveillance being felt as a threat, but as a chance to display oneself under the gaze of the camera." Ernst, however, also stresses that dataveillance is displacing surveillance.[63] More catastrophically, Paul Virilio claims that Webcams, which erase the interval between subject and image and thus make all meaning visible, are a "democratization of voyeurism on a planetary scale" that will lead to a visual crash.[64] More catastrophically still, Jean Baudrillard contends that reality media, by divorcing merit from accomplishment, will lead to species suicide, also known as radical democracy.[65] So from years of television and celebrity watching and do-it-yourself media production, being shot has become banal and celebrity democratized. Perhaps.

Webcams, or Democratizing Publicity

Webcams: supposedly live cameras, placed in homes and public places as well as on persons that transmit images over the Internet at varying frames per second. Webcam sites vary from the once-popular Cat Hospital, which featured a convalescing Frank the Cat, to endless pornographic sites that offer live chat, daily shows, and/or archives to paying members; from Webcams overlooking the Serengeti to those overlooking I-95 (see figures

---

61. Ursula Frohne, "'Screen Tests': Media Narcissism, Theatricality, and the Internalized Observer," in *CTRL [SPACE]: Rhetorics of Surveillance from Bentham to Big Brother*, eds. Thomas Y. Levin et al., 275, 262.

62. Peter Weibel, "Pleasure and the Panoptic Principle," in *CTRL [SPACE]*, eds. Levin et al., 208.

63. Wolfgang Ernst, "Beyond the Rhetoric of Perception: Surveillance as Cybernetics," in *CTRL [SPACE]*, eds. Levin et al., 461, 463.

64. Paul Virilio, *Open Sky*, trans. Julie Rose (London: Verso, 1997), 109.

65. Jean Baudrillard, "Telemorphosis," in Levin et al., *CTRL [SPACE]*, 480–485.

| Figure 5.7 |
Frank the Cat outside his cage ⟨http://www.cathospital.co.uk⟩

5.7–5.10).[66] Since many cam operators do not offer their cams, or are not on their site, 24/7, there are megaportal sites, such as camwhores.com, designed to imitate the control panel of a large surveillance operation (figure 5.11). On them, one can usually find at least one live cam at any given time. A system of perpetual rivalry determines inclusion onto such megasites. Two sites usually compete during a "test period": a site's popularity depends on nudity (if one will not get naked, one should be willing to do the outrageous, such as eat feces) and availability. Once on, one's stills are voted on constantly. "Click throughs," an important source of revenue, also induce perpetual rivalry: cam girls—even Goth cam girls with fuck-you attitudes—ask for them. The vast numbers of "cam whores" and the

66.   Cathospital.com garnered an award from Yahoo, thousands of hits daily, and worldwide media attention—much of this stemmed from it becoming "front-page" news on the bbc.co.uk Web site.

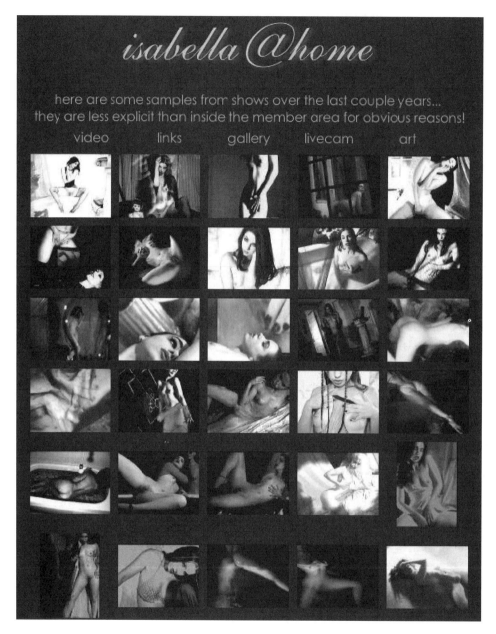

| Figure 5.8 |
Images taken from isabellacam.com, an art/porn site

| Figure 5.9 |
Pick of the day (January 19, 2003) from Africam

| Figure 5.10 |
Webcam located at 93/Tobin Bridge/Central Artery (Fleet Center) from www.smartraveler.com

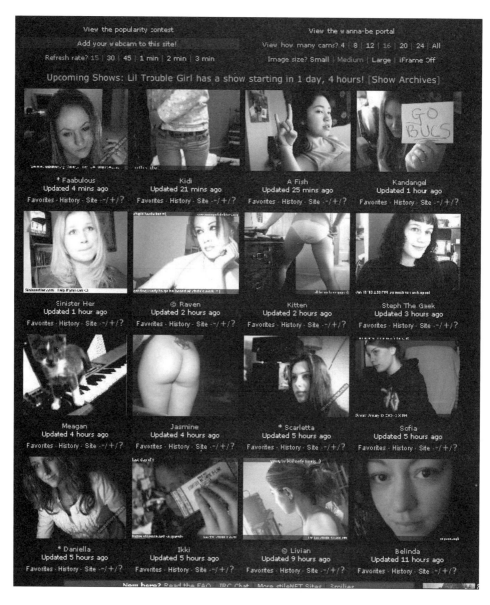

| Figure 5.11 |
Front portal to camwhores.com taken at 6:13 p.m., January 19, 2003

constant competitions make cam sites seem endless, but the inevitable overlap between megasites, as well as Web rings, belies this endlessness.

Webcams encapsulate perfectly the relationship between delusional control and freedom. Porn sites allow members to "control" the action during their live "shows"; nonpornographic sites, such as jennicam.com, spookycam.com, and anacam.com, do not encourage such "interactivity," but place cameras (which may or may not be on) in key locations within their operator's apartment/workplace.[67] They also usually keep a live journal (Ana Voog offers her paying members anagrams updated several times daily). These "girls" seemingly give up their privacy for a fleeting chance at celebrity. To many, these cam whores are perverse either because their actions are pornographic/erotic or because they are simply online, willingly suspending their human right to/need for privacy— something society only requires from (reluctant) celebrities and public figures. Such a willing suspension undermines the liberty that, according to Berlin, grounded the United States and other first world nations during the cold war. If perverse, however, the source is unclear: Does one become a cam whore because of some latent tendency toward exhibitionism, or does the camera itself induce perverse displays and desires for exposure? Is such "deviant behavior" the price of surveillance, a "contagion of a surveillance induced voyeurism and exhibitionism"?[68]

---

67. The difference between pornographic and nonpornographic Web sites arguably depends less on content and more on form, for nonpornographic Web sites do contain nudity and sexually explicit acts (necessary to prove their "liveness"). Pornographic Webcams tend to be on only during fixed periods of the day and feature more chances for "interactivity."

68. Brandon W. Joseph, "Nothing Special: Andy Warhol and the Rise of Surveillance," in *CTRL [SPACE]*, eds. Levin et al., 251. Not coincidentally, Internet porn sites perpetuate what the Diagnostic and Statistical Manual of Mental Disorders considers paraphilias, such as sadism and masochism, fetishism, voyeurism, and exhibitionism, although as argued in chapter 1, all pornography mimics voyeurism. According to Aiden of spookycam.com, quite a few cam operators suffer from panic disorders or depression, which is not surprising given that the ideal cam operator must be willing to stay indoors, like an injured cat, for long periods of time. Again Weibel in "Pleasure and the Panoptic Principle" argues that Webcams make social previously private and individual neuroses.

Importantly, from the Webcam operator's perspective or rhetoric, Webcams are all about choice (operators choose to be on camera) and freedom (freedom of expression and the freedom to experiment, although again this freedom, like Schreber's, does seem "this side of bureaucratization and human dignity"[69]). In their frequently asked questions sections, they invariably state that they are doing this because they can, because they don't mind, or because they—as Jennifer Ringley of jennicam emphasized—and not you, are conducting an experiment. Rather than "owning them," they own you; rather than merely being caught by surveillance cameras like everyone else, they choose when and how they are caught. According to Aiden of spookycam, they control the image we receive: they place the camera (and thus choose its blind spots); they turn it off and on. Through their artifice, users get a "false sense of knowing you."[70] Brought up screened (by television and film), users forget that their view is mediated, and that Webcams can be and indeed are faked: some operators on camwhores.com openly recycle their more pornographic images. Webcams may thus open operators' homes, but they do not expose them entirely. Operators' sites respond to and reveal the increasing irrelevance of liberal conceptions of privacy, and the move from private/public to open/closed.

Even the paid and ever-changing "girls" featured on voyeurdorm.com who seem unlikely defenders of control-cum-freedom, insist they are in control. "'What I really hate is that every guy demands that you take your top off, all the time,' says Tamra. 'It's like, I will take my top off *when* and *if* I want to. I used to say: Guys, go look up the definition of voyeur.'"[71] Interestingly, Tamra views the position of the object of the voyeur to be one of power. Vanessa Grigoriadis, who interviewed the women featured on voyeurdorm.com for nerve.com, sees Tamra's rhetoric as standard Webcam ideology:

---

69.   Friedrich Kittler, *Discourse Networks 1800/1900*, trans. Michael Metteer, with Chris Cullens (Stanford, CA: Stanford University Press, 1990), 303.

70.   Aiden, Unpublished interview (with Wendy Hui Kyong Chun).

71.   Quoted in Vanessa Grigoriadis, "I'm Seen, Therefore I Am," nerve.com, ⟨www.nerve.com/Dispatches/Grigoriadis/voyeurDorm/⟩ (accessed September 30, 1999).

If Tamra thinks there's something empowering about deciding when and when not to take her top off on camera, it's because she's been schooled in the Voyeur Dorm party line. The site, Hammill [the owner] maintains earnestly, is not a porn site, but a celebration of freedom of expression and sexual pride, a zone that merely records these young women in their natural and unashamed state. Excessive nudity, like lounging around buck-naked, is not encouraged by Hammill and there are no bonuses for it [there are bonuses for studying or other activities that invoke voyeurism and make the site coincide with its name]. The girls spend most of their waking hours in baby-Ts and miniskirts, typical mall-going regalia, or maybe the occasional sports bra. That said, they don't shy from exhibitionism: when they swim, it's usually topless, and often on chat they'll strip down at the insistence of the hundreds of rabid men on the other end of the modem, who then duly offer praise about the girls' beauty. Also, whether it speaks to their boredom, their "Gen Y" bi-curiousness, a sense of showmanship or a genuine desire for female affection, the girls do fool around with each other on camera in ever-shifting pairs.[72]

These "girls," empowered to perform their natural and unashamed state and their mainly newly discovered bisexuality for the camera, are not exhibitionists. According to the Diagnostic and Statistical Manual of Mental Disorders, exhibitionism entails the behavior or urge to expose one's genitals to an unsuspecting stranger: Web voyeur members are hardly unsuspecting or strangers (when signing up, they proffer their name and credit card number), and this exposure takes place within one's own house rather than in public.[73] Similarly, Webcam members are not voyeurs: the voyeur

---

72. Ibid.

73. Victor Burgin, in his analysis of jennicam, argues that Jennifer Ringley is not an exhibitionist because exhibitionism, according to Sigmund Freud, derives from voyeurism: Ringley is not interested in seeing ours, and popular diagnoses of Ringley as exhibitionist simply verify our own voyeurism. In contrast, Burgin, stressing Ringley's age, views jennicam as a substitute for the mother's gaze, which notices and approves of Jennifer: "Jenni is tottering around in her mother's shoes. Under the gaze of her mother she is investigating what it means to be a woman like her mother. That is to say, she is posing the question of female sexuality" ("Jenni's Room: Exhibitionism and Solitude," *Critical Inquiry* 27 [Fall 2000], 85). Accord-

gets off by watching unsuspecting strangers.[74] The relationship between the cam operator/model and the member can be quite familiar: members chat online with their cam person, send them e-mail, meet at certain public locations, send gifts, or buy things from their eBay site. Almost every cam operator has a "wish list"; Aiden, who describes herself as a mid-level player, received gifts, varying from CDs to laptops, weekly in 2003.

Through these interactions, the viewer feels less "creepy," for the viewer is acknowledged. This acknowledgment alleviates one of guilt, of the guilty pleasure of seeing without being seen, which one receives in excess of the contractual agreement. Geoffrey Batchen, in his analysis of photography as a guilty pleasure, remarks that, "far from being a marginal perversion, seeing without being seen has been a central tenet of the practice of photography throughout its history, a guilty pleasure thought to provide insights into life beyond the reach of the posed picture."[75] This pleasure, manifested in posed photographs shot as if unposed, conflates the unposed with "life." The intimacy between the watcher and his or her window combines this guilty pleasure with the relationship between pet owner and pet: someone or something is there for you—it may not be within view all the time but it is there, willing for you to look at or over it. It is something to love that does not talk or look back, that won't leave: it is a love like Nathaniel's love for Olympia in ETA Hoffmann's

---

ingly, the Internet becomes a space of "play" outside the individual, but not the external world. Burgin also asserts that Jenni's control over her presence and absence puts her in a position of (maternal) power in relation to her viewers. Linking this to Freud's analysis of his grandson's fort and da game in *Beyond the Pleasure Principle* (ed. and trans. James Strachey [New York: W. W. Norton, 1991]), Burgin notes that we viewers, like lonely adult children, "keep watch through their windows for Jenni to come home from work" ("Jenni's Room," 87). Burgin's argument that users wait quietly for Jenni to come home from work is insightful and complicates his claim that they are voyeurs. For if Ringley is not an exhibitionist, the viewers are not voyeurs.

74.   Freud's linking of voyeurism to cruelty, though in *Three Essays on the Theory of Sexuality*, ed. and trans. James Strachey (London: Hogarth Press, 1962), does seem to explain to some extent the hostility these women face.

75.   Geoffrey Batchen, "Guilty Pleasures," in *CTRL [SPACE]*, eds. Levin et al., 459.

The Sandman. Those who keep a window onto Ringley or Frank the Cat are looking for someone to look after, and not, as Victor Burgin argues, someone to come home to and have look after them. Webcams can epitomize surveillance as benevolence. When in 2001, "cam girl" Stacy Pershall attempted suicide live on the Internet, her watchers called the ambulance and were able to rescue her.

Many of the interactions between watcher and watchee, however, are hostile, and given the environment of constant competition for adoration, this hostility can be devastating. All cam operators receive threatening e-mails and constant demands to take off one's shirt; many of them collect the strangest of these e-mails in a "freaks" section. Grigoriadis, relayed the following dialogue:

Tree: The reporter girl better eat pussy tonight.
Rexx: The reporter girl does not look as hot as she did last night.[76]

These two statements—the demand to expose one's body and a derogatory comment about one's body—exemplify Webcams' mirroring of paranoid knowledge. The mirror image must expose itself, and its minor imperfections must be attacked (Lacan developed his conception of paranoid knowledge through his analysis of Aimee, the paranoid psychotic who attacked an actress, and his interpretation of the Papin sisters' case, in which two sisters, who were inseparable, viciously attacked and killed their employer and her daughter).

The viciousness of the attacks on these Webcam operators is remarkable. In July 2002, Jenni explained the disappearance and depression of her partner online:

Dex no longer wants to be on camera. Though he tolerates it as much as he can, there are some things about which he is just too self-conscious to share. I've been through it myself, so I know how hurtful it can be, when you cry on camera for instance, for people to email and call you weak, and a cry-baby, and berate your "self-pity" and "whining." I know it's intimidating when the

---

76.   Grigoriadis, "I'm Seen."

camera catches you naked, and people write in droves to say how fat and unattractive you are, to tell you to "get your fat ass off the internet."

To those of you who still follow the site who have written those emails—yes, there are still a few of you—I'd like to point out that the internet is not just for the thin or the pretty. The more I hear or read those sorts of things, the more convinced I become that what I do is meaningful.

Dex's spirit is less sturdy than mine, though. He knows he's overweight, so he doesn't want you to see him showering. I know I'm overweight, but whether I'm happy with the shape of my body, I want to make the point that I do not cease to exist. Not that I insist on being nude on camera, but that I refuse to avoid being nude on camera.[77]

Ringley here views her Webcam as democratizing the media, as revealing the fact that overweight people still exist, where still existing equals being in front of the camera. In this sense, Frohne's and others insistence that "I think, therefore I am" has been transformed to "I am seen, therefore I am" seems accurate.

Significantly, the insults directed toward these cam operators could easily be redirected to their writers. Daign of daign.com, a site dedicated to ridiculing cam girls, wrote in response to Stacy Pershall's attempted suicide:

Stacy is a cam girl. She's lived on cam for quite some time now, much like her cam girl role model friend Ana Voog. They thrive off of attention, like the rest of the cam girl community. Its a social substitute for some, for others its a necessity.

If I spent a couple years basing my life upon a relationship and the internet and I lost the relationship, I imagine I'd consider offing myself too.

After all, a webcams not gonna come off your monitor and give you a kiss, nor is it going to take you to dinner and let you know how beautiful you are.

Fact of the matter is. You cam girls are fucked up.

Afterall.

---

77. From the "journal" section of the now defunct jennicam, ⟨http://www.jennicam.com⟩ (accessed September 11, 2002).

———

What good are friends and lovers if they are pixels on a screen? Absolutely fucking nothing.

This event has only solidified my view on you fuckups. You might as well all get together in one location and commit mass suicide. Come on over to my house. I'll spike the punch. I'd like to see you all convulse and foam at the mouth on my disco dance floor. It'd be like an epileptic camwhore disco^{TM}.

Send me hatemail. Tell me how I'm wrong. Explain to me that even though it's the internet, these people still have feelings. Tell me that you have friends on the net and you love them.

You are all fucking delusional. Go outside and introduce yourself to someone. It's a nice thing for your soul, eh?[78]

Daign's accusation that these women (and men) are "fucked up" and delusional because they spend too much time on the Internet could easily be directed at Daign himself; to produce these infamous reviews, he had to spend a considerable amount of time online. Web operators' sense of control and the average users' are equally overinflated. Absolute control is on neither side of the camera. His accusation is also interestingly misdirected: Pershall's real-life lover left her; her friends on the Internet saved her life. The nastiness of his attack arguably stems from his own insecurities. As Daign's critique nicely reveals, by highlighting another's delusion, we create an outside that is even more profoundly delusional.

Significantly, the media, with the exception of a few online journals, did not respond to Stacy's suicide—the fake and extremely well-planned "our first time" Web site garnered far more attention. As the *Wired* commentator noted, general skepticism directed toward "live" Internet events affected media coverage: no matter how much we want to believe that Webcams are indexical, we are skeptical.[79] Commentators, who demonize the Internet's duplicity, ignore this skepticism, which infects everything on the Internet. Those who watch Webcams and believe, however, are

---

78. From the now defunct daign.com ⟨http://www.daign.com/girls/stacy/bio.html⟩ (accessed January 1, 2002).

79. See Julia Scheeres, "Dying for Attention?" *Wired News*, July 14, 2001, ⟨http://www.wired.com/news/culture/0,1284,45247,00.html?tw=wn_story_related⟩ (accessed September 3, 2004).

not simply naive, but are lured by the promise of authenticity, reinforced by these sites' "amateur" status.

Through refreshing Webcams, computers become live: no longer information processing machines, their wires appear truly connected elsewhere, their windows truly real-time. The computer screen changes without a mouse click. This surprise—this catching of movement—contrasts starkly to asynchronous Internet applications such as e-mail. Its gripping uneventfulness, its stationary camera, and its jerky refreshing also contrast sharply with television or film. One keeps watching a cam precisely because nothing happens. The plot of these Webcams, if there is one, is usually provided by accompanying live journal entries. The window does not need to be in focus: one does not need to watch it all the time. Rather, it is one window among many that one can check for changes—it is an opening. Webcams try to make the system visible, try to make fiber-optic networks transparent, as if there were a simple window, rather than an invisible and noisy system; as if there were such a thing as tele-presence rather than an intricate system of mediation and translation. Webcams promise to make computers prosthetic. As Thomas Campenella puts it, "Webcameras are a set of wired eyes, a digital extension of the human faculty of vision."[80] Yet as Wolfgang Ernst argues, dataveillance is not visual—the age of fiber optics makes the visual metaphoric. Emphasizing images or even a visual crash indicates a fascination with the visual that is surprisingly dissonant with the technology and with technology use, and furthers the intertwining of freedom with control.

Freedom

Thus far, I have outlined the ways in which control-freedom thrives on a paranoid knowledge that focuses on technological rather than political solutions and that relies on racial profiling. In this paranoid mind-set, freedom as autonomy—freedom from constraint, the sexual freedom to expose and reinvent ourselves, etc.—plays a key role in our simultaneous resignation to the reality of unfreedom and our delusion of freedom as

---

80.　Thomas Campanella, "Eden by Wire: Webcameras and the Telepresent Landscape" in *The Robot in the Garden: Telerobotics and Telepistemology in the Age of the Internet*, ed. Ken Goldberg (Cambridge: MIT Press, 2001), 23.

an intimate discourse grounding our independence and sexual freedom. Importantly, paranoia responds to a knowledge of a rottenness at the core of power—a knowledge that mechanisms of control sustain power, for which it seeks to compensate. Although Lacanian theories of paranoia have been critical to this analysis, in part because the rhetoric of control-freedom perpetuates paranoid knowledge as diagnosed by Lacan, these theories, in their insistence on the symbolic as the solution, are surprisingly blind to nonmetaphoric deprivations of freedom and to the role of the symbolic in colonialism. The increasing acceptance of paranoia also problematizes the normalizing function of language, pointing to the ways in which language—through its current privatization (DNA, programming languages, indefinitely prolonged copyright) and its non-phonetic variants—seems no longer to provide a firm "third relation."[81]

Freedom, however, cannot be reduced to control. Freedom exceeds, rather than complements, control. The ideological conflation of freedom with safety—the idea that we are only free when safe—defers freedom, and makes it an innocuous property of subjectivity. But freedom comes with no guarantees; it breaks bonds, enabling good *and* evil. The conflation of freedom and safety defers this violent opening, which Chela Sandoval and Frantz Fanon both see as a creative destruction, and seeks to make the future predictable. And every deferral of freedom also destroys it because freedom, as Jean-Luc Nancy argues, is "a fact" rather than an idea or a given; as such, it can only be experienced, where the experiencing of freedom is "a testing of something real."[82]

---

81.   Importantly, these non-phonetic languages are *not* natural languages, but constructed ones.

82.   Jean-Luc Nancy, *The Experience of Freedom*, trans. Bridget McDonald (Stanford, CA: Stanford University Press, 1993), 20. The fact of freedom does not mean that freedom is a given; rather, states Nancy, the fact of freedom, which is "radically 'established' without any establishing procedure being able to produce this fact as a theoretical object, is the fact of what is to be done ... it is the fact *that there is* something, to be done, or is even the fact that there is the *to be done*, or that there is the affair of existence. Freedom is factual in that it is the *affair* of existence.... Human beings are not born free in the same way that they are born with a brain; yet they are born, infinitely, to freedom" (20–21).

Nancy's notion of freedom as primary is an excellent starting point for understanding a freedom that exceeds control. Freedom, he observes, is not something we possess: it is not a thing, an idea, or an ideal. Freedom, or rather the "freely"—a generosity that precedes the possibility of any kind of possession—enables being to emerge in the first place. This freedom is *nothing*. For Nancy, it is a nothing that enables being and singularity through its withdrawal, through the spacing it enables through this withdrawal:

Freedom is that which spaces and singularizes—or which singularizes *itself*—because it is freedom of being in its withdrawal. Freedom "precedes" in the sense that being *cedes* before every birth to existence: it withdraws. Freedom *is* the withdrawal of being, but the withdrawal of being is the nothingness of this being, which is the being of freedom. This is why freedom *is not, but it frees being and frees from being*, all of which can be rewritten here as: *freedom withdraws being and gives relation.*[83]

Freedom is a spacing that constitutes existence. Freedom spaces in its withdrawal, and that there is something is the gift of this withdrawal; this withdrawal divides and joins, enabling existence, relation, and singularity. Freedom is not the lack of relation but the very possibility of relation, and thus of an existent as such (Nancy argues that one can view freedom as a cutting—or more properly, an opening—that enables an existent to exist). Freedom cannot be separated from fraternity or equality, for fraternity exists because we all share this nothing, and equality exists because we all measure ourselves against it. The history of freedom supports freedom as spacing: Nancy agrees with Hannah Arendt's contention that freedom was first understood as "the free space of movements and meetings; freedom as the external composition of trajectories and outward aspects, before being an internal disposition."[84] Fanon, not cited in Nancy's analysis (as be-

---

83.   Ibid., 68.

84.   Nancy writes that "free space cannot be opened through any subjective freedom. Free space is opened, freed, by the very fact that it is constituted or instituted as space *by* the trajectories and outward aspects of singularities that are thrown into existence. There is no space previously provided for displacement . . . , but there is

reft of details as Lacan's), similarly asserts that the "native" first dreams of freedom as action and aggression, as running faster than a motorcar.[85]

These movements outward pirate freedom or the freely, for freedom, which *is nothing*, can only be pirated.[86] The experience of freedom of movement, banal though it may seem, opens the possibility of freedom, for as Nancy points out,

freedom cannot be awarded, granted, or conceded according to a degree of maturity or some prior aptitude that would receive it. Freedom can only be *taken*; this is what the *revolutionary* tradition represents. Yet the taking of freedom means that freedom takes itself, that it has already received itself, from itself. No one begins *to be* free, but freedom *is* the beginning and endlessly remains the beginning.... The political does not primarily consist in the composition and dynamic of powers ..., but in the opening of a space. This space is opened by freedom—initial, inaugural, arising—and freedom there presents itself in action. Freedom does not produce anything, but only comes to produce itself there (it is not *poiesis*, but *praxis*), in the sense that an actor, in order to be the actor he is, produces himself on the stage.... Power has an origin, freedom is a beginning. Freedom does not cause coming-to-being, it is *an initiality of being*. Freedom is what is initially, or (singularly) *self-initiating* being. Freedom is the existence of the existent as such, which means that it is the initiality of its "setting into position."[87]

This emphasis on freedom as a beginning, as something that begins everything and every being, underscores the fact that no one is completely

a sharing and partitioning of origin in which singularities space apart and space their being-in-common" (ibid., 74).

85.   Franz Fanon, *The Wretched of the Earth*, trans. Constance Farrington (New York: Grove Press, 1968), 52.

86.   In Thomas Keenan's slightly different terminology, freedom allows for no stable identity, but only iteration, where we can understand this iteration not only in terms of Derrida's argument that iteration enables language but also iteration as the repeated act of spacing (*Fables of Responsibility: Aberrations and Predicaments in Ethics and Politics* [Stanford, CA: Stanford University Press, 1997], 91).

87.   Nancy, *Freedom*, 78.

alienated from freedom—no one can claim that freedom is "wasted" on a certain group of people because they are without an understanding of or appreciation for freedom. Indeed, Orlando Patterson has compellingly argued that freedom stemmed from the longing of slaves.[88] Significantly,

---

88.   Orlando Patterson, writing a social history of freedom as a *continuous* value in Western societies, sees freedom as a tripartite value that emerged from the experience of slavery. The three notes of Patterson's "chord" of freedom are personal, sovereignal, and civic freedom—where personal freedom is freedom from coercion and the ability to do as one pleases within the limits of that other person's desire to do the same; where sovereignal freedom is the power to act as one pleases regardless of another's desires; and civic freedom is the capacity of the adult members of a community to participate in the life and the governance of it (begging the question, What is a community and the relationship between community and governmental institutions?). Although notions of freedom have precursors in less complex and less "intellectually self-conscious people," Patterson maintains that freedom did not emerge as a value until the Greeks, for economic and political reasons (*Freedom: Freedom in the Making of Western Culture* [New York: Basic Books, 1991], 18). Patterson also argues that good and evil are tragically intertwined: personal liberty is the "noblest achievement of Western civilization," yet "no value has been more evil and socially corrosive in its consequences, inducing selfishness, alienation, the celebration of greed, and the dehumanizing disregard for the 'losers,' the little people who failed to make it" (403). Likewise, both Christianity and Nazism perpetuated sovereignal freedom; civic freedom produces democracy, but in both Athens and the United States, this democracy was "conceived in and fashioned by, the degradation of slaves and their descendents and the exclusion of women" (405). What the history of freedom teaches, then, is "out of evil cometh good" (405). Patterson's insistence on good and evil as intertwined rectifies conceptions of freedom as innocuously good, and his insistence that slavery generated freedom rectifies beliefs, like Rousseau's, that slaves despise freedom. These interventions force a reconsideration of cultural assumptions about the nobility and priority of freedom. Yet a cliché, "out of evil cometh good," does not adequately address the relationship between good and evil, and Patterson's history of the emergence of freedom as a value elides freedom as an experience or fact—of freedom as exceeding human values. Even if freedom did not emerge as a value until Athens, freedom existed as a fact well before then. Also, his elegant tripartite formulation limits differences in the value of freedom—from Athens to the present—to the stressing of one "note" over the other, begging the question, To what extent has the dissolution of formal slavery in most "Western" countries affected the value of freedom?

freedom as initiality, as power, makes freedom both good and evil; the first manifestation of freedom—the withdrawal of being and the furious unleashing—is wickedness. Evil absolutely ruins good—it destroys good before it can occur; it is freedom unleashed against itself.[89] Freedom necessitates a decision: a decision for good is a decision for finitude, a decision to hold back its possibility for devastation; a decision for evil is the letting go of this furious devastation.

Nancy's insistence on freedom as a fact, as something that precedes us, and therefore as something that is foreign to no one and something that no one can grant to another is important, but his formulation does leave many questions unanswered. If evil ruins absolutely, how can good prevail? Must liberation wait for a (liberal) decision for good? What if we "let" ourselves be exposed and others do not? What if others "let" us be exposed? How do we deal with inequality? Arguably, Nancy's formulation screens inequality through its emphasis on space and debt. Nancy remarks, "There is no space previously provided for displacement ..., but there is a sharing and partitioning of origin in which singularities space apart and space their being-in-common."[90] This notion of shared nondisplaced space, however, grounds the founding of the "new world" as frontier and the erasure of native peoples. Power as the opening of space is also the displacement of others; pirates pirate—they plunder another's fortune. The dream of an ever-giving, never-displacing well of generosity uncannily resonates with the Internet as infinite capitalism. As well, the notion of a debt owed by all to existence flattens differences and inadvertently resonates with the current perpetuation of slavery through debt. Regardless, Nancy's insistence that freedom enables sharing and relation, precisely because it is nothing, helps us to imagine a nonautonomous freedom beyond control, and his insistence that freedom can enable both good and evil helps us assess the dangers of freedom. Freedom cannot be reduced to something innocuous. Freedom entails a decision for life or death.

---

89. Nancy, *Freedom*, 126.

90. Ibid., 74.

This choice is more pressing than ever because biopower—the power over life—has been made symbolic (minus the symbolics), if not semantic. The power of language, which Fanon and Keenan saw as making possible freedom and colonization (Fanon after all did write), increasingly stems from its nonphonetic variants. Because genes have been extracted from DNA as software from hardware, our bodies can now be read for our genealogy and history. For African American descendants of slavery, this control as *contreroule* affords an opportunity to build a genealogy against their history's erasure. Our history is now our bodies; the truth we hide is not our sexuality but our genome. DNA, combined with its twin software, is giving sexuality a decidedly nondiscursive function: namely, the recombination of genes. Genetics, though, has also made the importance of sexuality to race transparent: races are "breeding populations." Hence, biopower, the reproduction of life, continues, with sex as just one tool in its arsenal.

These silent languages end humans' singular claim to language, if not speech. According to Friedrich Kittler, language is moving beyond humans toward machines: "Data flows once confined to books and later to records and films are disappearing into black holes and boxes that, as artificial intelligences, are bidding us farewell on their way to nameless high commands. In this situation we are left only with reminiscences, that is to say, with stories."[91] Humans, having lost the power to write (machines now write themselves, enabled by reflexive negative feedback control systems), can only tell stories as information machines bypass their human so-called inventors. More strongly, Kittler claims, "bees are projectiles, and humans, cruise missiles. One is given objective data on angles and distances by a dance, the other, a command of free will. Computers operating IF-THEN commands are therefore machine subjects. Electronics, a tube monster since Bletchley Park, replaces discourse, and programmability replaces free will."[92] Kittler's assessments may seem hyperbolic and, for someone who declared that "there is no software," oddly

---

91.　Friedrich Kittler, *Film, Gramophone, Typewriter*, trans. Geoffrey Winthrop-Young (Stanford, CA: Stanford University Press, 1999), xxxix.

92.　Ibid., 259.

accepting of software's status as a language. If software is a language—if computers operate if-then commands—it is only because software has been constructed as such for the human operator. If the genome is considered a language, it is because of a similar construction, and these constructions are the end of a cybernetic dream based on a technology that perpetuates master-and-slave relations, that reduces freedom to control, language to programs and commands. This paranoid mirroring between machines and animals, paradoxically enabled by and displacing of language, points toward a future dominated by the redundancy of digital manipulation.

These new languages may not allow for polysemy—meaning can only be opened by rewriting the languages and their compilers—but the future remains open. We still play a role in the creation of our machines and their languages, and through our technologies—through our always compromised using—we can imagine and move toward a different future. To do so, however, we must engage all four layers of networks together and refuse easy assertions of freedom at one level that cover over unfreedom at another. To face this future and seize the democratic potential of fiber-optic networks, we must reject current understandings of freedom that make it into a gated community writ large. We must explore the democratic potential of communications technologies—a potential that stems from our vulnerabilities rather than our control. And we must face and seize freedom with determination rather than fear and alibis.

EPILOGUE

Fiber-optic networks. A literalization of "enlightenment" and the rays of Schreber's God. A system of light that creates a network of networks by proliferating female plugs and points of contact. Fiber-optic networks spread the light and conflate message with medium, so that we no longer see the light through our glass tubes. What we do see via, if not through, them seems delusional and hallucinatory, supposedly consensually so. Although we do not mistake personal hallucination for reality, developers hope that one day everyone, not just paranoid schizophrenics, will be unable to distinguish between pictured humans and real ones. Fiber-optic networks have exploded and circulated "deviant" enjoyments and cross-gender fantasies. They demand constant examination and response; one must enjoy or think. There can be no rest, or to be more precise, even when we rest, our machines communicate without us. Rather than our or its constant activity being contrary to the order of the world, however, such activity enables communication. Fiber-optic networks process information without knowledge. Real-time knowledge is possible and thrilling, but expensive and cumbersome. Fiber-optic networks comprise a synthesis of biology and machine technology; they assume the "stuff" of the human mind can be stored, and they dream of immortality through the separation of body from memory. In other words, they assume that our nerves can live without the body, or more modestly, that memories can be stored prosthetically and then sifted.[1] They also dream of virtual reality, of

---

1.　This notion of nerves as immortal is remarkably similar to the cybernetic extrapolations of Norbert Wiener and the perhaps schizophrenic writings of William Burroughs. For Burroughs, the destruction of the human body may be evolutionarily necessary for interplanetary travel. More immediately, Microsoft is

being able to "picture" something that someone else who "rides" one's eyes receives.[2]

According to Avital Ronell, the early twentieth-century schizophrenic had telephone receivers and electric currents running along its body; the early twenty-first-century paranoid views these receivers as data jacks and concentrates on voltage signifiers rather than running current.[3] *According to Friedrich Kittler, writers simulated madness in 1900; in 2003, reality simulates madness, and everyone—especially Kittler—seems to conflate fact with fiction.*[4] *In 2003, all those with female (computer) plugs are encouraged to change gender and publish their secrets via live journals/blogs in order to become authors/objects for observation, if there is a difference between the two. This explosion of online confessions, which can be understood as the democratization of the talking cure (no money is exchanged and no expertise is necessary), compensates for the increasing localization and medication of mental anguish. This "democratization" of the analysand-analyst relation denotes both the triumph and the failure of psychoanalysis. Although medical psychiatry now predominates and, according to Žižek, the talking cure is no longer effective in analysis, people still talk and still use their body to signify.*[5] *Sex is still a secret that needs at least two people to decipher. Psychoanalysis has become diffused throughout society: it has become co-opted in a do-it-yourself craze.*

Importantly, in 1903 Schreber's vision was singular, and paranoia paranormal—Schreber lost his civil rights and was incarcerated in an insane asylum; Schreber also considered himself singular, a messiah. In the age of computerization, as Intel CEO Andy Grove so eloquently put it, "only the paranoid sur-

---

working on a "memory system" that will allow you to sort through all your digital memories, solving the "photobox problem."

2.   This notion is uncannily similar to William Gibson's notion of simstim.

3.   See Avital Ronell, *The Telephone Book: Technology—Schizophrenia—Electric Speech* (Lincoln: University of Nebraska Press, 1989), 109–121, 263–264.

4.   See Friedrich Kittler, *Discourse Networks 1800/1900*, trans. Michael Metteer, with Chris Cullens (Palo Alto, CA: Stanford University Press, 1990), 307.

5.   See Slavoj Žižek, *The Ticklish Subject: The Absent Centre of Political Ontology* (London: Verso, 1999).

vive."[6] *Also, whereas Schreber fought to be released from tutelage (and thus freed from his wife's financial decisions), we are now all encouraged to be free— to wander about and believe that we are beyond tutelage (that is, as long as we take our medication). This difference, this freedom of movement—a freedom akin to the freedom of a commodity in the so-called free market—is the key difference between our situation and Schreber's. Marx once contended that bourgeois freedom was in essence the freedom of capital and that the relations between humans were perceived as the relations between commodities. Now, we not only fetishize commodities but also actively emulate them—and this emulation or transparency, which arguably rends the veil of ideology, has not made commodity fetishism, or ideology, any less powerful.*

Paranoia does not respond to an overwhelming, all-seeing power but rather to a power found to be lacking—rotten and inadequate, always decaying. This decay, this inability to simply "be," seemingly comes from elsewhere. Although obsessed with contamination by others, paranoia stems from those who are already jacked in, who have already transgressed into the domain of ludertum, who have already in some way identified with those others. And once jacked in, it seems impossible to cut the connection or make it fruitfully reproductive, no matter how repeatedly we think voluptuous thoughts or purify those "tested" souls inhabiting us.

Rather than bearing witness to the twinning of liberty and discipline, however, we are testifying to a different crisis that we, like Schreber, are experiencing as sexuality—and as a freedom, in Kittler's words, "this side of bureaucratization and human dignity."[7] *If then we were in a realm of discipline and liberty—one in which the localization of mental disorders could be considered a form of soul murder—we are now in one of control and freedom. In a realm in which mental disorders are treated as chemical imbalances and the human body treated as a decodable control system, this freedom is experienced as a freedom from discipline. Control-freedom enables a far finer resolution than liberty-discipline, but not as fine as its propaganda: it is a system in which simulations are offered as evidence*

6.   See Andrew Grove, *Only the Paranoid Survive* (New York: Currency, Doubleday, 1996).

7.   Friedrich Kittler, *Discourse Networks 1800/1900*, 303.

References

Abbas, Ackbar. *Hong Kong: Culture and the Politics of Disappearance*. Minneapolis: University of Minnesota Press, 1997.

Allstetter, Rob. "Entertainment: Japanese Videos Get Animated Interest." *Detroit News*, January 17, 1996, J1.

Althusser, Louis. "Ideology and Ideological State Apparatuses (Notes towards an Investigation)." In *Lenin and Philosophy and Other Essays*. Translated by Ben Brewster. New York: Monthly Review Press, 1971.

American Civil Liberties Union. "Drawing a Blank." ⟨http://archive.aclu.org/issues/privacy/drawing_blank.pdf⟩ (accessed May 1, 2004).

Anime Turnpike. ⟨http://www.anipike.com⟩ (accessed May 15, 1999).

Arai, Andrea G. "The 'Wild Child' of 1990s Japan." *SAQ* 99, no. 4 (2000): 841–863.

Argyle, Katie, and Rob Shields. "Is There a Body on the Net?" In *Cultures of Internet: Virtual Spaces, Real Histories, Living Bodies*, edited by Rob Shields, 58–69. London: Sage Publications, 1996.

Asian Nudes. ⟨http://www.asiannudes.com/tour1.html⟩ (accessed April 1, 1999).

Bagdikian, Ben. *The Media Monopoly*. 5th ed. Boston: Beacon Press, 1997.

Balibar, Etienne, and Immanuel Wallerstein. *Race, Nation, Class: Ambiguous Identities*. Translated by Chris Turner. London: Routledge, Chapman and Hall, 1991.

Balsamo, Anne. "Forms of Technological Embodiment: Reading the Body in Contemporary Culture." In *Cyberspace/Cyberbodies/Cyberpunk: Cultures of*

Technological Embodiment, edited by Michael Featherstone and Roger Burrows, 215–237. London: Sage Publications, 1995.

Barlow, John Perry. "Africa Rising." ⟨http://www.wired.com/wired/archive/6.01/barlow.html⟩ (accessed May 1, 1999).

———. "A Declaration of the Independence of Cyberspace." ⟨http://www.salon1999.com/08/features/declaration.html⟩ (accessed May 1, 1999).

Batchen, Geoffrey. *Burning with Desire: The Conception of Photography.* Cambridge: MIT Press, 1997.

———. "Guilty Pleasures." In *CTRL [SPACE]: Rhetorics of Surveillance from Bentham to Big Brother*, edited by Thomas Y. Levin et al., 446–459. Cambridge: MIT Press, 2002.

Baudrillard, Jean. *The Ecstasy of Communication.* Translated by Bernard and Caroline Schutze. Brooklyn, NY: Semiotext(e), 1988.

———. "Telemorphosis." In *CTRL [SPACE]: Rhetorics of Surveillance from Bentham to Big Brother*, edited by Thomas Y. Levin et al., 480–485. Cambridge: MIT Press, 2002.

Baudry, John-Louis. "The Apparatus: Metapsychological Approaches to the Impression of Reality in the Cinema." In *Narrative, Apparatus, Ideology: A Film Theory Reader*, edited by Philip Rosen, 299–318. New York: Columbia University Press, 1986.

Benjamin, Walter. "The Work of Art in the Age of Mechanical Reproduction." In *Illuminations: Essays and Reflections*, translated by Harry Zohn, 217–251. New York: Schocken Books, 1968.

———. *The Arcades Project.* Translated by Howard Eilan and Kevin McLaughlin. Cambridge: Harvard University Press, 1999.

Bentham, Jeremy. "The Panopticon; or, the Inspection-House." In *The Panopticon Writings*, edited by Miran Božovič, 29–114. London: Verso, 1995.

Berlant, Lauren. "National Brands/National Body: *Imitation of Life.*" In *The Phantom Public Sphere*, edited by Bruce Robbins, 173–209. Minneapolis: University of Minnesota Press, 1997.

Berlin, Isaiah. *Two Concepts of Liberty: An Inaugural Lecture Delivered before the University of Oxford on 31 October 1958.* Oxford: Clarendon Press, 1958.

Binelli, Mark. "Large Eyes Blazing, Anime Offers Exotic Views." *Atlanta Constitution*, October 27, 1995, P10.

Blade Runner. Directed by Ridley Scott. Burbank, CA: The Ladd Company, 1982.

Booker, M. Keith. "Technology, History, and the Postmodern Imagination: The Cyberpunk Fiction of William Gibson." *Arizona Quarterly* 50, no. 4 (Winter 1994): 63–87.

Brande, David. "The Business of Cyberpunk: Symbolic Economy and Ideology in William Gibson." In *Virtual Realities and Their Discontents*, edited by Robert Markley, 79–106. Baltimore, MD: Johns Hopkins University Press, 1996.

"'Bring It Down' Was about a Car, Students' Lawyer Says." CNN.com. ⟨http://www.cnn.com/2002/US/09/15/fla.terror.students/⟩ (accessed September 13, 2003).

Brook, James, and Iain A. Boal, eds. *Resisting the Virtual Life: The Culture and Politics of Information*. San Francisco: City Lights, 1995.

Buckley, Sandra. "'Penguin in Bondage': A Graphic Tale of Japanese Comic Books." In *Technoculture*, edited by Andrew Ross and Constance Penley, 163–195. Minneapolis: University of Minnesota Press, 1991.

Buell, Frederick. "Nationalist Postnationalism: Globalist Discourse in Contemporary American Culture." *American Quarterly* 50, no. 3 (September 1998): 548–591.

Burgin, Victor. "Jenni's Room: Exhibitionism and Solitude." *Critical Inquiry* 27 (Fall 2000): 77–89.

Burroughs, William. *The Adding Machine: Selected Essays*. New York: Henry Holt, 1986.

Butler, Judith. *Bodies That Matter: On the Discursive Limits of "Sex."* New York: Routledge, 1993.

———. "Endangered/Endangering: Schematic Racism and White Paranoia." In *Reading Rodney King, Reading Urban Uprising*, ed. Robert Gooding-Williams. New York: Routledge, 1993.

———. *Excitable Speech: A Politics of the Performative*. New York: Routledge, 1997.

Butler, Octavia. *Parable of the Sower*. New York: Warner Books, 1993.

———. *Mind of My Minds*. New York: Warner Books, 1977.

———. *Parable of the Talents*. New York: Warner Books, 2000.

Cadigan, Pat. *Synners*. New York: Bantam Books, 1991.

Campanella, Thomas. "Eden by Wire: Webcameras and the Telepresent Landscape." In *The Robot in the Garden: Telerobotics and Telepistemology in the Age of the Internet*, edited by Ken Goldberg, 22–46. Cambridge, MA: MIT Press, 2001.

Catholic Agency for Overseas Development. *Clean Up Your Computer*. ⟨http://www.cafod.org.uk/policy_and_analysis/policy_papers/clean_up_your _computer_report⟩ (accessed May 1, 2004).

Cherny, Lynn, and Elizabeth Reba Weiss, eds. *Wired Women*. Seattle: Seal Press, 1996.

Chesher, Chris. "The Ontology of Digital Domains." In *Virtual Politics: Identity and Community in Cyberspace*, edited by David Holmes, 79–93. London: Sage Publications, 1997.

Cheskin Research. *The Digital World of the US Hispanic*. April 2000.

Chow, Rey. *Primitive Passions: Visuality, Sexuality, Ethnography, and Contemporary Chinese Cinema*. New York: Columbia University Press, 1995.

———. *Ethics after Idealism: Theory, Culture, Ethnicity, Reading*. Bloomington: Indiana University Press, 1998.

———. "Gender and Representation." In *Feminist Consequences*, edited by Elisabeth Bronfen and Misha Kavka. New York: Columbia University Press, 2000.

Chun, Wendy Hui Kyong. "Introduction: Did Somebody Say New Media." In *New Media, Old Media: A History and Theory Reader*, edited by Wendy Hui Kyong Chun and Thomas Keenan, 1–10. New York: Routledge, 2005.

———. "On Software, or the Persistence of Visual Knowledge." *Grey room* 18 (winter 2005): 26–51.

Crary, Jonathan. *Suspensions of Perception: Attention, Spectacle, and Modern Culture*. Cambridge, MA: MIT Press, 1999.

Crenshaw, Kimberlé. "Color Blindness, History, and the Law." In *The House That Race Built: Black Americans, U.S. Terrain*, edited by Wahneema Lubiano, 280–288. New York: Pantheon, 1997.

Csicsery-Ronay, Istvan, Jr. "Antimancer: Cybernetics and Art in Gibson's *Count Zero*." *Science Fiction Studies* 22 (1995): 63–86.

"Cybersex: Policing Pornography on the Internet." *ABC Nightline*, June 27, 1995.

daign.com. 〈http://www.daign.com/girls/stacy/bio.html〉 (accessed January 1, 2002).

Darnton, Robert. *The Literary Underground of the Old Regime*. Cambridge: Harvard University Press, 1982.

———. *The Forbidden Best-Sellers of Pre-Revolutionary France*. New York: W. W. Norton, 1995.

Dean, Jodi. *Publicity's Secret: How Technoculture Capitalizes on Democracy*. Ithaca, NY: Cornell University Press, 2002.

de Certeau, Michel. *The Practice of Everyday Life*. Translated by Steven Rendall. Berkeley: University of California Press, 1984.

Deleuze, Gilles. "Postscript on Control Societies." In *CTRL [SPACE]: Rhetorics of Surveillance from Bentham to Big Brother*, edited by Thomas Y. Levin et al., 317–321. Cambridge: MIT Press, 2002.

Deleuze, Gilles, and Felix Guattari. *Anti-Oedipus: Capitalism and Schizophrenia*. Translated by Robert Hurley, Mark Seem, and Helen R. Lane. Minneapolis: University of Minnesota Press, 1983.

Derrida, Jacques. "Signature Event Context." In *Limited Inc.*, translated by Samuel Weber and Jeffrey Mehlman, 1–23. Evanston, IL: Northwestern University Press, 1988.

Dershowitz, Alan M. "Why Fear National ID Cards?" *New York Times*. 〈http://www.nytimes.com/2001/10/13/opinion/13DERS.html〉 (accessed October 14, 2001).

Dery, Mark. "Black to the Future: Interviews with Samuel R. Delany, Greg Tate, and Tricia Rose." *SAQ* 92, no. 4 (Fall 1993): 735–778.

———. *Escape Velocity*. New York: Grove Press, 1996.

Deutsch, Rosalyn. *Evictions: Art and Spatial Politics*. Cambridge: MIT Press, 1996.

de Zwann, Victoria. "Rethinking the Slipstream: Kathy Acker Reads *Neuromancer*." *Science Fiction Studies* 24 (1997): 459–470.

Dibell, Julian. *My Tiny Life: Crime and Passion in a Virtual World*. New York: Owl Books, 1998.

Dienst, Richard. *Still Life in Real Time: Theory after Television*. Durham, NC: Duke University Press, 1994.

Diffie, Whitfield, and Susan Landau. *Privacy on the Line: The Politics of Wiretapping and Encryption*. Cambridge: MIT Press, 1998.

Doane, Mary Ann. *The Desire to Desire: The Woman's Film of the 1940s*. Bloomington: Indiana Uiversity Press, 1987.

———. "Information, Crisis, and Catastrophe." In *Logics of Television: Essays in Cultural Criticism*, edited by Patricia Mellencamp, 222–239. Bloomington: Indiana University Press, 1990.

Eisenstein, Sergei. *Eisenstein on Disney*. Translated and edited by Jay Leyda. London: Methuen, 1988.

Electronic Frontier Foundation. "'Censorship: Internet Censorship Legislation and Regulation, 1998' Archive." ⟨http://www.eff.org/pub/Censorship/Internet_censorship_bills/1998_bills/⟩ (accessed March 4, 1999).

Elmer-Dewitt, Philip. "On a Screen Near You, Cyberporn: A New Study Shows How Pervasive and Wild It Really Is." *Time*, July 3, 1995, 38–45.

Enemy of the State. Directed by Tony Scott. Los Angeles: Jerry Bruckheimer Productions (Touchstone), 1998.

Eriksen, Inge. "The Aesthetics of Cyberpunk." *Foundation 53* (Fall 1991): 36–46.

Ernst, Wolfgang. "Beyond the Rhetoric of Perception: Surveillance as Cybernetics." In *CTRL [SPACE]: Rhetorics of Surveillance from Bentham to Big Brother*, edited by Thomas Y. Levin et al., 460–463. Cambridge: MIT Press, 2002.

———. "Dis/continuities: Does the *Archive* Become Metaphorical in Multimedia Space?" In *New Media, Old Media: A History and Theory Reader*, edited by Wendy Hui Kyong Chun and Thomas Keenan, 105–123. New York: Routledge, 2005.

Evenson, Laura. "Cyberbabe Takes on Tokyo in 'Ghost'; Tough, Topless Cartoon Heroine." *San Francisco Chronicle*, April 12, 1996, D3.

Everett, Anna. "The Revolution Will Be Digitalized: Afrocentricity and the Digital Public Sphere." *Social Text* 20, no. 2 (Summer 2002): 125–246.

Fabian, Johannes. *Time and the Other: How Anthropology Makes Its Object*. New York: Columbia University Press, 1983.

Fabijancic, Tony. "Space and Power: Nineteenth-Century Urban Practice and Gibson's Cyberworld." *Mosaic* 32, no. 1 (March 1999): 105–139.

"Facial Recognition Technology May Screen for Terrorists." CBSnews.com. ⟨http://www.cbsnews.com/stories/2002/01/31/health/printable327398.shtml⟩ (accessed September 13, 2003).

Fanon, Frantz. *Black Skin, White Masks*. Translated by Charles Lam Markmann. New York: Grove Press, 1967.

———. *The Wretched of the Earth*. Translated by Constance Farrington. New York: Grove Press, 1968.

Farnell, Ross. "Posthuman Topologies: William Gibson's 'Architexture' in *Virtual Light* and *Idoru*." *Science Fiction Studies* 25, no. 3 (1998): 459–480.

Featherstone, Michael, and Roger Burrows, eds. *Cyberspace/Cyberbodies/ Cyberpunk: Cultures of Technological Embodiment*. London: Sage Publications, 1995.

Federal Reserve Bank of Richmond. *Seeing the Light*. ⟨http://www.rich.frb .org/pubs/regionfocus/summer03/light.html⟩ (accessed May 1, 2004).

Feuer, Jane. "The Concept of Live Television: Ontology as Ideology." In *Regarding Television: Critical Approaches—An Anthology*, edited by E. Ann Kaplan, 12–22. Frederick, MD: University Publications of America, 1983.

"Fighting Terror: Government Overhaul President's Address; 'Freedom and Fear Are at War.'" *Boston Globe*, June 7, 2002, 3rd ed., A37.

Findlen, Paula. "Humanism, Politics, and Pornography in Renaissance Italy." In *The Invention of Pornography: Obscenity and the Origins of Modernity, 1500– 1800*, edited by Lynn Hunt, 49–108. New York: Zone Books, 1993.

Foucault, Michel. *Discipline and Punish: The Birth of the Prison*. Translated by Alan Sheridan. New York: Vintage Books, 1978a.

———. *The History of Sexuality, Volume 1: An Introduction*. Translated by Robert Hurley. New York: Vintage Books, 1978b.

———. "The Eye of Power." In *Power/Knowledge: Selected Interviews and Other Writings, 1972–1977*, translated by Colin Gordon, 146–165. New York: Pantheon Books, 1980a.

———. "Two Lectures." In *Power/Knowledge: Selected Interviews and Other Writings, 1972–1977*, edited by Colin Gordon, 78–108. New York: Pantheon Books, 1980b.

———. "Of Other Spaces." Translated by A. M. Sheridan Smith. *diacritics* (Spring 1986): 22–27.

Fox, Kit. "Interconnectivity: Three Interviews with the Staff of *Lain*." *Animerica* 7, no. 9 (October 1999): 27–29.

Franklin, Gene F., J. David Powell, and Abbas Emami-Naeini. *Feedback Control of Dynamic Systems*. Reading, MA: Addison-Wesley, 1988.

Frappier-Mazur, Lucienne. "The Truth and the Obscene Word in Eighteenth-Century French Pornography." In *The Invention of Pornography: Obscenity and the Origins of Modernity, 1500–1800*, edited by Lynn Hunt, 203–221. New York: Zone Books, 1993.

Fraser, Nancy. "Rethinking the Public Sphere: A Contribution to the Critique of Actually Existing Democracy." In *The Phantom Public Sphere*, edited by Bruce Robbins, 1–32. Minneapolis: University of Minnesota Press, 1997.

Freud, Sigmund. *Three Essays on the Theory of Sexuality*. Edited and translated by James Strachey. London: Hogarth Press, 1962.

———. "Psychoanalytic Notes upon an Autobiographical Account of a Case of Paranoia (Dementia Paranoides)." In *Three Case Histories*, 83–160. New York: Collier Books, 1963.

———. *Beyond the Pleasure Principle*. Edited and translated by James Strachey. New York: W. W. Norton, 1991.

Frohne, Ursula. "'Screen Tests': Media Narcissism, Theatricality, and the Internalized Observer." In *CTRL [SPACE]: Rhetorics of Surveillance from Bentham to Big Brother*, edited by Thomas Y. Levin et al., 252–277. Cambridge: MIT Press, 2002.

Fuller, Matthew. "The Mouths of the Thames." ⟨http://www.tate.org.uk/webart/mongrel/home/faqs/ns.htm⟩ (accessed February 26, 2001).

Fuss, Diana. *Identification Papers*. New York: Routledge, 1995.

Galloway, Alexander. "Protocol, or, How Control Exists after Decentralization." *Rethinking Marxism* 13, nos. 3/4 (Fall/Winter 2001): 81–88.

———. "Tactical Media and Conflicting Diagrams." *Nettime*. ⟨http://amsterdam.nettime.org/Lists-Archives/nettime-l-0301/msg00047.html⟩ (accessed September 13, 2003).

———. "Institutionalization of Computer Protocols," *Nettime*. ⟨http://amsterdam.nettime.org/Lists-Archives/nettime-l-0301/msg00052.html⟩ (accessed May 1, 2004).

Gates, Bill. *The Road Ahead*. New York: Viking, 1995.

Ghost in the Shell (Kôkaku kidôtai). Directed by Mamoru Oshii. Tokyo: Production IG, 1995.

Gibson, William. *Neuromancer*. New York: Ace Books, 1984.

———. *Count Zero*. New York: Ace Books, 1986.

———. *Mona Lisa Overdrive*. Toronto: Bantam Books, 1988.

———. *Idoru*. New York: G. P. Putnam's Sons, 1996.

———. "Interview with Addicted to Noise." ⟨http://www.addict.com/issues/2.10/html/hifi/Cover_Story/⟩ (accessed February 1, 2000).

———. *Pattern Recognition*. New York: Putnam, 2003.

Gingrich, Newt. *To Renew America*. New York: HarperCollins, 1995.

Ginsburg, Elaine K., ed. *"Passing" and the Fictions of Identity*. Durham, NC: Duke University Press, 1996.

Godwin, Mike. "Journoporn: Dissection of the *Time* Scandal." *Hotwired*. ⟨http://hotwired.wired.com/special/pornscare/godwin.html⟩ (accessed May 1, 2004).

Gonzalez, Jennifer. "The Appended Subject: Race and Identity as Digital Assemblage." In *Race in Cyberspace*, edited by Beth Kolko et al., 27–50. New York: Routledge, 2000.

Gourley, David, and Brian Totty. *HTTP: The Definitive Guide*. New York: O'Reilly, 2002.

Green, David, and David Kidwell. "Federal Sources Say Terrorism Threat by Three Students Was a Hoax." *Miami Herald Online Edition* (September 13, 2003). ⟨http://www.miami.com/mld/miami/4068519.htm⟩ (accessed October 1, 2003).

Greenwald, Jeff. "Wiring Africa." ⟨http://www.wired.com/wired/archive/2.06/africa.html⟩ (accessed May 1, 1999).

Grigoriadis, Vanessa. "I'm Seen, Therefore I Am." nerve.com. ⟨www.nerve.com/Dispatches/Grigoriadis/voyeurDorm/⟩ (accessed September 30, 1999).

Grigsby, Mary. "Sailormoon: Manga (Comics) and Anime (Cartoon) Super-heroine Meets Barbie; Global Entertainment Commodity Comes to the United States." *Journal of Popular Culture* 32, no. 1 (Summer 1998): 59–80.

Grove, Andrew. *Only the Paranoid Survive*. New York: Currency, Doubleday, 1996.

Gunning, Tom. "From Kaleidoscope to the X-Ray: Urban Spectatorship, Poe, Benjamin, and Traffic in Souls (1913)." *Wide Angle* 19, no. 4 (1997): 25–63.

Habermas, Jürgen. *The Structural Transformation of the Public Sphere: An Inquiry into the Category of Bourgeois Society*. Translated by Thomas Burger. Cambridge: MIT Press, 1991.

Hackers. Directed by: Iain Softley. Los Angeles: United Artists, 1995.

Hale, Grace Elizabeth. *Making Whiteness: The Culture of Segregation in the South, 1890–1940*. New York: Pantheon Books, 1998.

Haraway, Donna. *Simians, Cyborgs, and Women: The Reinvention of Nature*. New York: Routledge, 1991.

Hardt, Michael, and Antonio Negri. *Empire*. Cambridge: Harvard University Press, 2000.

Hartman, Saidiya. *Scenes of Subjection: Terror, Slavery, and Self-Making in Nineteenth-Century America*. New York: Oxford University Press, 1997.

Hayles, N. Katherine. *Writing Machines*. Cambridge: MIT Press, 2002.

Healy, Dave. "Cyberspace and Place: The Internet as Middle Landscape on the Electronic Frontier." In *Internet Culture*, edited by David Porter, 55–68. New York: Routledge, 1997.

Hebert, Wray. "Paranoia: Fearful Delusions." *New York Times Magazine*. March 19, 1989, 62.

Hecht, Jeff. *City of Light: The Story of Fiber Optics*. New York: Oxford University Press, 1999.

Heidegger, Martin. "The Age of the World Picture." In *Electronic Culture: Technology and Visual Representation*, edited by Timothy Druckrey, 47–61. New York: Aperture, 1996.

Hibbard, Allen, ed. *Conversations with William S. Burroughs*. Jackson: University Press of Mississippi, 1999.

Hill, Logan, and Thuy Linh Nguyen. "Asian Artists Make Porn Sites Work for Them: Nude Japanese Schoolgirls! Lotus Blossoms! Radical Feminists?" *Village Voice.* ⟨http://www.villagevoice.com/issues/0134/hill.php⟩ (accessed August 25, 2001).

Hoffmann, Donna, and Thomas Novak. "A Detailed Analysis of the Conceptual, Logical, and Methodological Flaws in the Article: 'Marketing Pornography on the Information Superhighway.'" ⟨http://elab.vanderbilt.edu/research/topics/cyberporn/rimm.review.htm⟩ (accessed May 1, 2004).

Hollinger, Veronica. "Cybernetic Deconstructions: Cyberpunk and Postmodernism." *Mosaic* 23, no. 2 (Spring 1990): 29–43.

Horibuchi, Seiji. "Interview with Rumiko Takahashi." In *Anime Interviews: The First Five Years of ANIMERICA, ANIME AND MANGA MONTHLY (1992–1997)*, edited by Trish Ledoux, 16–23. San Francisco: Cadence Books, 1997.

Horn, Carl Gustav. "Interview with Mamoru Oshii." In *Anime Interviews: The First Five Years ANIMERICA, ANIME, AND MANGA MONTHLY (1992–1997)*, edited by Trish Ledoux, 134–141. San Francisco: Cadence Books, 1997.

Huffman, Kathy Rae. "Video, Networks, and Architecture." In *Electronic Culture: Technology and Visual Representation*, edited by Timothy Druckrey, 200–207. New York: Aperture, 1996.

Hunt, Craig. *TCP/IP Network Administration.* 2nd ed. New York: O'Reilly, 1997.

Hunt, Lynn, ed. *The Invention of Pornography: Obscenity and the Origins of Modernity, 1500–1800.* New York: Zone Books, 1993.

Hwang, David Henry. *M. Butterfly.* New York: Penguin, 1989.

IEEE. "About the Society." IEEE Control Systems Society (CSS). ⟨http://www.ieeecss.org/about/ABOUTindex.html⟩ (accessed September 13, 2003).

"Interview with Mamoru Oshii." *ALLES.* ⟨http://www.express.co.jp/ALLES/6/oshii2.html⟩ (accessed May 1, 1999).

Ivy, Marilyn. "Revenge and Recapitation in Recessionary Japan." *SAQ* 99, no. 4 (2000): 819–840.

Jameson, Fredric. "Progress versus Utopia; or, Can We Imagine the Future?" *Science Fiction Studies* 9, no. 2 (July 1982): 147–158.

References

———. *Postmodernism, or The Cultural Logic of Late Capitalism*. Durham, NC: Duke University Press, 1991.

———. *Signatures of the Visible*. New York: Routledge, 1992.

jennicam. ⟨http://www.jennicam.com⟩ (accessed September, 2002).

Jeremijenko, Natalie. "Dialogue with a Monologue: Voice Chips and the Products of Abstract Speech." ⟨http://cat.nyu.edu/natalie/VoiceChips.pdf⟩ (accessed September 13, 2002).

Johnston, John. "Computer Fictions: Narratives of Machinic Phylum." *Journal of the Fantastic in the Arts* 8, no. 4 (Fall 1997): 443–463.

Joseph, Brandon W. "Nothing Special: Andy Warhol and the Rise of Surveillance." In *CTRL [SPACE]: Rhetorics of Surveillance from Bentham to Big Brother*, ed. Thomas Y. Levin et al., 237–251. Cambridge: MIT Press, 2002.

Kakoudaki, Despina. "Pinup and Cyborg: Exaggerated Gender and Artificial Intelligence." In *Future Females, the Next Generation*, edited by Marleen S. Barr, 165–195. Boulder, CO: Rowman and Littlefield, 2000.

Kant, Immanuel. "An Answer to the Question: What Is Enlightenment?" In *What is Enlightenment? Eighteenth-Century Answers and Twentieth Century Questions*, edited and translated by James Schmidt, 58–64. Berkeley: University of California Press, 1996.

Katti, Christian. "'Systematically' Observing Surveillance: Paradoxes of Observation according to Niklas Luhmann's Systems Theory." In *CTRL [SPACE]: Rhetorics of Surveillance from Bentham to Big Brother*, ed. Thomas Y. Levin et al., 50–63. Cambridge: MIT Press, 2002.

Kawash, Samira. *Dislocating the Color Line: Identity, Hybridity, and Singularity in African-American Literature*. Stanford, CA: Stanford University Press, 1997.

Keating, Dan. "New Voting Systems Assailed." *Washington Post*, March 28, 2003, A12.

Keenan, Thomas. *Fables of Responsibility: Aberrations and Predicaments in Ethics and Politics*. Stanford, CA: Stanford University Press, 1997a.

———. "Windows: Of Vulnerability." In *The Phantom Public Sphere*, edited by Bruce Robbins, 121–141. Minneapolis: University of Minnesota Press, 1997b.

Kendrick, Michelle. "Cyberspace and the Technological Real." In *Virtual Realities and Their Discontents*, edited by Robert Markley, 143–160. Baltimore, MD: Johns Hopkins University Press, 1996.

Kipnis, Laura. *Bound and Gagged: Pornography and the Politics of Fantasy in America*. Durham, NC: Duke University Press, 1999.

Kittler, Friedrich. *Discourse Networks 1800/1900*. Translated by Michael Metteer, with Chris Cullens. Foreword by David E. Wellbery. Stanford, CA: Stanford University Press, 1990.

———. *Gramophone, Film, Typewriter*. Translated and with an introduction by Geoffrey Winthrop-Young and Michael Wutz. Stanford, CA: Stanford University Press, 1999.

———. "There Is No Software." ⟨http://www.ctheory.net/textfile.asp?pick =74⟩ (accessed August 1, 2004).

———. "Cold War Networks or Kaiserstr. 2, Neubabelsberg." In *New Media, Old Media: A History and Theory Reader*. Edited by Wendy Hui Kyong Chun and Thomas Keenan, 181–186. New York: Routledge, 2005.

Kosofsky Sedgwick, Eve. *Epistemology of the Closet*. Berkeley: University of California Press, 1990.

La Bare, Joshua. "The Future: 'Wrapped … in That Mysterious Japanese Way.'" *Science Fiction Studies* 27, no. 1 (March 2000): 22–48.

Lacan, Jacques. *The Psychoses, 1955–1956: The Seminar of Jacques Lacan, Book III*. Edited by Jacques-Alain Miller. Translated by Russel Grigg. New York: W. W. Norton, 1993.

Laclau, Ernesto, and Chantal Mouffe. *Hegemony and Socialist Strategy: Towards a Radical Democratic Politics*. 2nd ed. London: Verso, 2001.

Lamarre, Thomas. "From Animation to *Anime*: Drawing Movements and Moving Drawings." *Japan Forum* 14, no. 2 (2002): 329–367.

LambdaMOO Papers. ⟨ftp://parc.xerox.com/pub/MOO/Papers⟩ (accessed April 1, 1996).

Landow, George. *Hypertext: The Convergence of Contemporary Critical Theory and Technology*. Baltimore, MD: Johns Hopkins University Press, 1992.

Lane, Frederick S. *Obscene Profits: The Entrepreneurs of Pornography in the Cyber Age*. New York: Routledge, 2000.

Lazarowitz, Elizabeth. "COLUMN ONE: Beyond 'Speed Racer': Japanese Animation Has Exploded in Popularity Worldwide; Creators of Such New-Generation Superheroes as a Female Cyber-Cop Hope to Cash in on TV Shows, Videos and Comic Books." *Los Angeles Times*, December 3, 1996, 1.

Ledoux, Trish. "Interview with Masamune Shirow." In *Anime Interviews: The First Five Years of ANIMERICA, ANIME, AND MANGA MONTHLY (1992–1997)*, edited by Trish Ledoux, 38–45. San Francisco: Cadence Books, 1997.

Lee, Edmund. "Anime of the People." *Village Voice*, April 9, 1996, 6.

Lefort, Claude. *Democracy and Political Theory*. Translated by David Macey. Minneapolis: University of Minnesota Press, 1988.

Lerman, Laurence. "Anime Vids Get Euro-Friendly." *Variety*, June 24, 1996, 103.

Lessig, Lawrence. *Code: And Other Laws of Cyberspace*. New York: Basic Books, 1999.

———. *The Future of Ideas: The Fate of the Commons in a Connected World*. New York: Random House, 2001.

Levi, Antonia. *Samurai from Outer Space: Understanding Japanese Animation*. Chicago: Open Court, 1996.

Levin, Thomas Y. "Rhetoric of the Temporal Index: Surveillant Narration and the Cinema of Real Time." In *CTRL [SPACE]: Rhetorics of Surveillance from Bentham to Big Brother*, edited by Thomas Levin et al., 578–593. Cambridge: MIT Press, 2002.

Levy, Steven. "Tech's Double-Edged Sword." *Newsweek*, September 24, 2001. ⟨http://amsterdam.nettime.org/Lists-Archives/nettime-l-0110/msg00042.html⟩ (accessed May 1, 2004).

Lindberg, Kathyne V. "Prosthetic Mnemonics and Prophylactic Politics: William Gibson among the Subjectivity Mechanisms." *Boundary 2* (Summer 1996): 45–83.

Loshin, Pete. *TCP/IP Clearly Explained*. 3rd ed. San Diego, CA: Academic Press, 1999.

Lovink, Geert. *Dark Fiber: Tracking Critical Internet Culture*. Cambridge, MA: MIT Press, 2002.

Lyotard, Jean-François. *The Postmodern Condition: A Report on Knowledge*. Translated by Geoff Bennington and Brian Massumi. Minneapolis: University of Minnesota Press, 1984.

MacKinnon, Catharine. *Only Words*. Cambridge: Harvard University Press, 1993.

———. "Vindication and Resistance: A Response to the Carnegie Mellon Study of Pornography in Cyberspace." *Georgetown Law Review* 93, no. 5 (June 1995): 1959–1967.

Maney, Kevin. "Bin Laden's Messages Could Be Hiding in Plain Sight." *USA Today Online*. ⟨http://www.usatoday.com/tech/columnist/2001/12/19/maney.htm⟩ (accessed September 12, 2003).

Manovich, Lev. *The Language of New Media*. Cambridge: MIT Press, 2001.

Marx, Karl. *Capital*. Translated by Ben Fowkes. Vol. 1. New York: Penguin Books with New Left Review, 1976.

Marx, Karl, and Friedrich Engels. *Manifesto of the Communist Party*. Peking: Foreign Languages Press, 1975.

McCaffery, Larry, ed. *Storming the Reality Studio: A Casebook of Cyberpunk and Postmodern Science Fiction*. Durham: Duke University Press, 1991.

MCI. "Anthem." New York: Messner Vetere Berger McNamee Schmetterer/ Euro RSCG, 1997.

McLuhan, Marshall. *Understanding Media*. Cambridge: MIT Press, 1994.

McPherson, Tara. "Reload: Liveness, Mobility and the Web." In *The Visual Culture Reader, 2nd ed.*, edited by Nicholas Mirzoeff, 458–470. New York: Routledge, 2002.

Mill, John Stuart. *On Liberty*. Chicago: Gateway Editions, n.d.

Ministry of Foreign Affairs of Japan and the United Nations University. "Global Partnership for Peace, Progress, and Prosperity: A Message from Africa." ⟨http://www.unu.edu/africa/oau2000.html⟩ (accessed February 26, 2001).

Moglen, Eben. "Anarchism Triumphant: Free Software and the Death of Copyright." *First Monday* 4, no. 8 (August 2, 1999). ⟨http://firstmonday.org/ issues/issue4_8/moglen/index.html⟩ (accessed May 1, 2004).

Mongrel. Natural Selection Star Sites. ⟨http://www.mongrelx.org/Project/ Natural/StarSites/starsites.html⟩ (accessed February 26, 2001).

———. Projects. ⟨http://www.mongrelx.org/Project/projects.html⟩ (accessed February 26, 2001).

Monnet, Livia. "Towards the Feminine Sublime, or the Story of 'a Twinkling Monad, Shape-Shifting across Dimension': Intermediality, Fantasy, and

Special Effects in Cyberpunk Film and Animation." *Japan Forum* 14, no. 2 (2002): 225–268.

Morley, David, and Kevin Robins. "Techno-Orientalism: Futures, Foreigners, and Phobias." *New Formations* (Spring 1992): 136–156.

Morse, Margaret. *Virtualities: Television, Media Art, and Cyberculture*. Bloomington: Indiana University Press, 1998.

Moylan, Tom. "Global Economy, Local Texts: Utopian/Dystopian Tension in William Gibson's Cyberpunk Trilogy." *Minnesota Review* 43–44 (1995): 182–197.

Mulvey, Laura. "Visual Pleasure and Narrative Cinema." In *Visual and Other Pleasures*, 14–26. Bloomington: Indiana University Press, 1989.

Nakamura, Lisa. *Cybertypes: Race, Ethnicity, and Identity on the Internet*. New York: Routledge, 2003.

Nancy, Jean-Luc. *The Experience of Freedom*. Translated by Bridget McDonald. Stanford, CA: Stanford University Press, 1993.

National Telecommunications and Information Administration. *Falling through the Net II*, July 1998. ⟨http://www.ntia.doc.gov/ntiahome/net2/falling .html⟩ (accessed February 26, 2001).

———. *Falling through the Net: Toward Digital Inclusion*, October 2000. ⟨http://www.ntia.doc.gov/ntiahome/fttn00/falling.htm⟩ (accessed February 26, 2001).

Nead, Lynda. *The Female Nude: Art, Obscenity, and Sexuality*. New York: Routledge, 1992.

———. "'Above the Pulp-Line': The Cultural Significance of Erotic Art." In *Dirty Looks: Women, Pornography, Power*, edited by Pamela Church Gibson and Roma Gibson, 144–155. London: British Film Institute, 1993.

Negroponte, Nicolas. "The Third Shall Be First." ⟨http://www.wired.com/ wired/archive/6.01/negroponte.html⟩ (accessed May 1, 1999).

The Net. Directed by Irwin Winkler. New York: Columbia Pictures (Sony), 1995.

Newitz, Annalee. "Magical Girls and Atomic Bomb Sperm: Japanese Animation in America." *Film Quarterly* 49, no. 1 (Fall 1995): 2–15.

Nunes, Mark. "What Space Is Cyberspace? The Internet and Virtuality." In *Virtual Politics: Identity and Community in Cyberspace*, edited by David Holmes, 163–178. London: Sage Publications, 1997.

Odzer, Cleo. *Virtual Spaces: Sex and the CyberCitizen*. New York: Berkley Books, 1997.

Office of the Whitehouse. *National Security Strategy of the United States of America*. ⟨http://www.whitehouse.gov/nsc/nss.html⟩ (accessed October 1, 2003).

Olsen, Lance. "Virtual Termites: A Hypotextual Technomutant Explo(it)ration of William Gibson and the Electronic Beyond(s)." *Style* 29, no. 2 (Summer 1995): 287–313.

Palais, Joseph C. *Fiber Optic Communications*. 4th ed. Upper Saddle River, NJ: Prentice Hall, 1998.

Parpis, Eleftheria. "Anime Action: Japanimation Is Edgy and Cool—and Shops Love It." *Adweek*, December 14, 1998, 18–20.

Patterson, Orlando. *Freedom: Freedom in the Making of Western Culture*. New York: Basic Books, 1991.

Piper, Adrian. "Passing for White, Passing for Black." *Transition* 58 (1993): 4–32.

Pointon, Susan. "Transcultural Orgasm as Apocalypse: *Urotsukidoji: The Legend of the Overfiend*." *Wide Angle* 19, no. 3 (July 1997): 41–63.

Porter, David, ed. *Internet Culture*. New York: Routledge, 1997.

Poster, Mark. *What's the Matter with the Internet?* Minneapolis: University of Minnesota Press, 2001.

Rajagopal, Arvind. "Imperceptible Perceptions in Our Technological Modernity." In *New Media, Old Media: A History and Theory Reader*, edited by Wendy Hui Kyong Chun and Thomas Keenan, 275–284. New York: Routledge, 2005.

Rheingold, Howard. *The Virtual Community: Homesteading on the Electronic Frontier*. Reading, MA: Addison-Wesley, 1993.

———. "A Slice of Life in My Virtual Community." In *Big Dummies' Guide to the Internet: A Round Trip through Global Networks, Life in Cyberspace, and Everything*, textinfo edition 1.02 (September 1993). ⟨http://www.hcc.hawaii.edu/bdgtti/bdgtti-1.02_18.html#SEC191⟩ (accessed June 1, 1999).

Rimm, Marty. "Marketing Pornography on the Information Superhighway: A Survey of 917,410 Images, Descriptions, Short Stories, and Animations Downloaded 8.5 Million Times by Consumers in over Forty Countries, Provinces, and Territories." *Georgetown Law Journal* 83, no. 5 (June 1995): 1849–1934.

Robbins, Bruce, ed. *The Phantom Public Sphere*. Minneapolis: University of Minnesota Press, 1997.

Robins, Kevin. "Cyberspace and the World We Live In." In *Cyberspace/Cyberbodies/Cyberpunk: Cultures of Technological Embodiment*, edited by Michael Featherstone and Roger Burrow, 135–155. London: Sage Publications, 1991.

Robinson, Amy. "It Takes One to Know One: Passing and Communities of Common Interest." *Critical Inquiry* 20, no. 4 (1994): 715–736.

Romney, Jonathan. "Manga for All Seasons: A Festival at the NFT Shows There Is More to Japan's Cult Anime Movies Than Misogyny and Apocalyptic Animation." *Guardian*, May 4, 1995, T.015.

Ronell, Avital. *The Telephone Book: Technology—Schizophrenia—Electric Speech*. Lincoln: University of Nebraska Press, 1989.

———. *Crack Wars: Literature, Addiction, Mania*. Lincoln: University of Nebraska Press, 1992.

Rosen, Jeffrey. "Being Watched: A Cautionary Tale for a New Age of Surveillance. *New York Times Magazine*. ⟨http://www.nytimes.com/2001/10/07/magazine/07SURVEILLANCE.html⟩ (accessed October 10, 2001).

Rosenthal, Pam. "Jacked-In: Fordism, Cyberspace, and Cyberpunk." *Socialist Review* (Spring 1991): 79–103.

Ross, Farnell. "Posthuman Topologies: William Gibson's 'Architexture' in *Virtual Light* and *Idoru*." *Science Fiction Studies* 25, no. 3 (1998): 459–480.

Rouse, Roger. "Thinking through Transnationalism: Notes on the Cultural Politics of Class Relations in the Contemporary United States." *Public Culture* 7 (Winter 1995): 353–402.

Saco, Diana. *Cybering Democracy: Public Space and the Internet*. Minneapolis: University of Minnesota Press, 2002.

Said, Edward. *Orientalism*. New York: Vintage Books, 1978.

Sakai, Naoki. "'You Asians': On the Historical Role of the West and Asia Binary." *SAQ* 99, no. 4 (2000): 789–817.

Sandoval, Chela. *Methodology of the Oppressed*. Minneapolis: University of Minnesota Press, 2000.

Santner, Eric. *My Own Private Germany: Daniel Paul Schreber's Secret History of Modernity*. Princeton, NJ: Princeton University Press, 1996.

Scheeres, Julia. "Dying for Attention?" *Wired News* (July 14, 2001). ⟨http://www.wired.com/news/culture/0,1284,45247,00.html?tw=wn_story _related⟩ (accessed September 3, 2004).

Schmitt, Ronald. "Mythology and Technology: The Novels of William Gibson," *Extrapolation* 34, no. 1 (1993): 64–78.

Schodt, Frederik. *Dreamland Japan: Writings on Modern Manga*. Berkeley, CA: Stone Bridge Press, 1996.

Scholes, Robert. *Structural Fabulation: An Essay on Fiction of the Future*, University of Notre Dame Ward-Phillips Lectures in English Language and Literature, vol. 7. Notre Dame, IN: University of Notre Dame, 1975.

Schreber, Daniel Paul. *Memoirs of My Nervous Illness*. Translated by Ida Macalpine and Richard A. Hunter. Cambridge: Harvard University Press, 1988.

Schroeder, Randy. "Determinacy, Indeterminacy, and the Romantic in William Gibson." *Science Fiction Studies* 21 (1994): 155–163.

———. "Neu-Criticizing William Gibson." *Extrapolation* 35, no. 4 (1994): 330–341.

Securiteam.com. "Step-by-Step Guide to DNA Poisoning." ⟨http://www .securiteam.com/securitynews/Domain-Hijacking-A-step-by-step-guide.html⟩ (accessed September 1, 2004).

Segal, Lynne. "Does Pornography Cause Violence? The Search for Evidence." In *Dirty Looks: Women, Pornography, Power*, edited by Pamela Church Gibson and Roma Gibson, 5–21. London: British Film Institute, 1993.

Segler, Jeffrey L. "The Right Thing: When the Boss Tumbles." *New York Times*, June 20, 1999, money and business/financial desk section. ⟨http://www .nytimes.com⟩ (document 41670).

Sekula, Allan. "The Body and the Archive." *October* 39 (1986): 3–64.

Senft, Theresa M. "Introduction: Performing the Digital Body—A Ghost Story." *Women and Performance*. ⟨http://www.echonyc.com/~women/Issue17/ introduction.htm⟩ (accessed June 8, 1999).

Shields, Rob, ed. *Cultures of Internet: Virtual Spaces, Real Histories, Living Bodies*. London: Sage Publications, 1996.

Shirow, Masamune. *Ghost in the Shell*. Translated by Frederik Schodt and Toren Smith. Milwaukie, OR: Dark Horse Comics, 1995.

Silverman, Kaja. *Male Subjectivity at the Margins*. New York: Routledge, 1992.

Simon, Bruce. "White-Blindness." In *The Social Construction of Race and Ethnicity in the United States*, edited by Joan Ferrante and Prince Brown Jr., 496–502. New York: Longman, 1998.

Smith, Andy. "Okay … Where's the Cyberporn?" *The Providence Journal-Bulletin* (July 16, 1995): 1E.

"Smithsonian without Walls." ⟨http://www.si.edu/revealingthings/⟩ (accessed May 1, 2004).

Sniffer FAQ. ⟨http://www.robertgralpubs/sniffing-tag.html⟩ (accessed September 1, 2003).

Sobchack, Vivian. "Beating the Meat/Surviving the Text, or How to Get out of This Century Alive." In *Cyberspace/Cyberbodies/Cyberpunk: Cultures of Technological Embodiment*, edited by Michael Featherstone and Roger Burrows, 205–214. London: Sage Publications, 1995.

Solomon, Charles. "For Kids, a 'Magical' Sampling of Japanese Animated Stories; Movies: UCLA Archive Caters to Growing Interest in Anime with Screenings of Features and Shorts." *Los Angeles Times*, January 8, 1999, 10.

Spigel, Lynn. *Make Room for TV: Television and the Family Ideal in Postwar America*. Chicago: University of Chicago Press, 1992.

Spillers, Hortense. "Mama's Baby, Papa's Maybe." In *Within the Circle: An Anthology of African American Literary Criticism from the Harlem Renaissance to the Present*, edited by Angelyn Mitchell, 454–481. Durham, NC: Duke University Press, 1994.

Sponsler, Claire. "William Gibson and the Death of Cyberpunk." In *Modes of the Fantastic*, edited by Robert A. Latham and Robert A. Collins, Westport, CT: Greenwood, 1991.

———. "Cyberpunk and the Dilemmas of Postmodern Narrative: The Example of William Gibson." *Contemporary Literature* 33, no. 4 (1992): 624–644.

Stephenson, Neal. *Snow Crash*. New York: Bantam Books, 1992.

References

————. "Mother Earth Mother Board." *Wired* 4, no. 12. ⟨http://www.wired .com/4.12/ffglass.html⟩ (accessed January 1, 1999).

Stockton, Sharon. "'The Self Regained': Cyberpunk's Retreat to the Imperium." *Contemporary Literature* 36, no. 4 (1995): 588–612.

Stone, Allucquère Rosanne. *The War of Desire and Technology at the Close of the Mechanical Age*. Cambridge: MIT Press, 1995.

Suvin, Darko. "On the Poetics of the Science Fiction Genre." *College English* 34 (1972): 372–382.

————. "On Gibson and Cyberpunk SF." *Foundations* 46 (Fall 1989): 40–51.

"Three Men Freed after Being Held for Hours in Florida over 'Alarming' Comments." *Jefferson City New Tribune*. ⟨http://newstribune.com/stories/ 091402/wor_0914020027.asp⟩ (accessed September 13, 2003).

"Three Suspects Played Stupid Joke, Feds Say." *NBC6 South Florida*. ⟨http:// www.nbc6.net/news/1667232/detail.html⟩ (accessed September 13, 2003).

Tocqueville, Alexis de. *Democracy in America*. Translated by Henry Reeve. New York: Knopf, 1994.

Trebilcock, Bob. "Child Molesters on the Internet." *Redbook*, April 1, 1997, 100–107.

Turk, Matthew A., and Alex P. Pentland. "Face Recognition Using Eigenfaces." In *Proceedings of the IEEE Conference on Computer Vision and Pattern Recognition*, 586–591. Maui, Hawaii, 1991.

Turkle, Sherry. *Life on the Screen: Identity in the Age of the Internet*. New York: Simon and Schuster, 1995.

Ueno, Toshiyo. "Japanimation and Techno-Orientalism." ⟨http://www.t0.or .at/ueno/japan.htm⟩ (accessed May 1, 1999).

"Understanding Biometrics: Face Recognition Technology." Identix. ⟨http:// www.identix.com/newsroom/news_biometrics_face.html⟩ (accessed September 13, 2003).

United Nations. "Development and International Cooperation in the Twenty-first Century: The Role of Information Technology in the Context of a Knowledge-Based Global Economy—Report of the Secretary-General," May 2000. ⟨http://www.un.org/documents/ecosoc/docs/2000/e2000-52.pdf⟩ (accessed February 26, 2001).

———. "Report of the Meeting of the High-Level Panel of Experts on Information and Communication Technology," May 2000. ⟨http://www.un.org/documents/ecosoc/docs/2000/e2000-55.pdf⟩ (accessed February 26, 2001).

U.S. Congress. House. *Cyberporn: Protecting Our Children from the Back Alleys of the Internet.* 104th Cong. Washington, DC: U.S. Government Printing Office, 1995.

———. *Report of the House of Representatives, Child Online Protection Act.* 105th Cong., 2d sess. ⟨http://www.epic.org/free_speech/censorship/hr3783-report.html⟩ (accessed January, 2001).

U.S. Court of Appeals for the Third Circuit. *American Civil Liberties Union v. Reno II*, no. 99–1324.

U.S. Department of Justice. *Department of Justice Brief (Reno v. ACLU), Filed with the Supreme Court on January 21, 1997.* ⟨http://www.ciec.org/SC_appeal/970121_DOJ_brief.html⟩ (accessed May 21, 1998).

———. *The Use of Computers in the Sexual Exploitation of Children.* Office of Justice Programs, 1999.

U.S. District Court for the Eastern District of Pennsylvania. *American Civil Liberties Union v. Reno.* Civ. A. No. 96–0963. ⟨http://www.eff.org/pub/Censorship/Exon_bill/HTML/960612_aclu_v_reno_decision.html⟩ (accessed May 21, 1998).

U.S. Supreme Court. *Concurrence by O'Connor/Rehnquist: Reno v. American Civil Liberties Union et al.* ⟨http://www.ciec.org/SC_appeal/concurrence.html⟩ (accessed September 19, 1997).

———. *Supreme Court Opinion (no 96–511): Reno v. American Civil Liberties Union et al.* ⟨http://www.ciec.org/SC_appeal/opinion.html⟩ (accessed September 19, 1997).

———. *Syllabus of Supreme Court Decision in Reno v. ACLU.* ⟨http://www.ciec.org/SC_appeal/syllabus.html⟩ (accessed September 19, 1997).

Virilio, Paul. *Open Sky.* Translated by Julie Rose. London: Verso, 1997.

———. "The Visual Crash." In *CTRL [SPACE]: Rhetorics of Surveillance from Bentham to Big Brother*, edited by Thomas Y. Levin et al., 108–113. Cambridge: MIT Press, 2002.

von Neumann, John. *First Draft of a Report on the EDVAC.* ⟨www.cs.colorado.edu/~zathras/csci3155/EDVAC_vonNeumann.pdf⟩ (accessed September 12, 2003).

Warner, Michael. "The Mass Public and the Mass Subject." In *The Phantom Public Sphere*, edited by Bruce Robbins, 234–256. Minneapolis: University of Minnesota Press, 1997.

Weber, Thomas. "The X Files: For Those Who Scoff at Internet Commerce, Here's a Hot Market: Raking in the Millions, Sex Sites Use Old-Fashioned Porn and Cutting-Edge Tech—Lessons for the Mainstream." *Wall Street Journal*, August 20, 1997, A1.

Weheliye, Alex. "Feenin: Posthuman Voices in Contemporary Black Popular Music." *Social Text* 20, no. 2 (Summer 2002): 21–47.

Weibel, Peter. "Pleasure and the Panoptic Principle." In *CTRL [SPACE]: Rhetorics of Surveillance from Bentham to Big Brother*, edited by Thomas Y. Levin et al., 206–223. Cambridge: MIT Press, 2002.

Weinstone, Ann. "Welcome to the Pharmacy: Addiction, Transcendence, and Virtual Reality." *Diacritics* 27, no. 3 (1997): 77–89.

"Why a Penguin?" ⟨http://www.linnx.org/info/penguin.html⟩ (accessed January 1, 2004).

Wiener, Norbert. *The Human Use of Human Beings: Cybernetics and Society*. New York: Da Capo Press, 1954.

———. *Cybernetics, or Control and Communications in the Animal and the Machine*. 2nd ed. Cambridge: MIT Press, 1961.

Williams, Linda. *Hardcore: Power, Pleasure, and the "Frenzy of the Visible."* Berkeley: University of California Press, 1999.

Williams, Patricia J. *The Alchemy of Race and Rights: Diary of a Law Professor*. Cambridge: Harvard University Press, 1991.

Women and Performance 17. ⟨http://www.echonyc.com/~women/Issue17⟩ (accessed June 8, 1999).

Wray, Stefan. "The Electronic Disturbance Theater and Electronic Civil Disobedience." ⟨http://www.thing.net/~rdom/ecd/EDTECD.html⟩ (accessed January 1, 2003).

Yekwai, Dimela. ⟨http://www.mongrelx.org/Project/Natural/Venus⟩ (accessed February 1, 2001).

Yoda, Tomiko. "The Rise and Fall of Maternal Society: Gender, Labor, and Capital in Contemporary Japan." *SAQ* 99, no. 4 (2000): 865–902.

Yudice, George. "We Are *Not* the World." *Social Text* 31–32 (1992): 202–216.

XXXAsians. ⟨http://www.xxxasians.com⟩ (accessed February 9, 2001).

Žižek, Slavoj. *The Sublime Object of Ideology*. London: Verso, 1989.

———. *The Plague of Fantasies*. London: Verso, 1997.

———. *The Ticklish Subject: The Absent Centre of Political Ontology*. London: Verso, 1999.

Index